THE MALAY ARCHIPELAGO

VOLUME I.

BY

ALFRED RUSSEL WALLACE

Copyright © 2013 Read Books Ltd.
This book is copyright and may not be
reproduced or copied in any way without
the express permission of the publisher in writing

British Library Cataloguing-in-Publication Data
A catalogue record for this book is available from the
British Library

THE MALAY ARCHIPELAGO, VOLUME I. (of II.)

By Alfred Russel Wallace

The land of the orang-utan, and the bird of paradise.
A narrative of travel, with sketches of man and nature.

To CHARLES DARWIN,

AUTHOR OF "THE ORIGIN OF SPECIES,"

I dedicate this book,
Not only as a token of personal esteem and friendship
But also To express my deep admiration
For His genius and his works.

Alfred Russel Wallace

Alfred Russel Wallace was born on 8th January 1823 in the village of Llanbadoc, in Monmouthshire, Wales.

At the age of five, Wallace's family moved to Hertford where he later enrolled at Hertford Grammar School. He was educated there until financial difficulties forced his family to withdraw him in 1836. He then boarded with his older brother John before becoming an apprentice to his eldest brother, William, a surveyor. He worked for William for six years until the business declined due to difficult economic conditions.

After a brief period of unemployment, he was hired as a master at the Collegiate School in Leicester to teach drawing, map-making, and surveying. During this time he met the entomologist Henry Bates who inspired Wallace to begin collecting insects. He and bates continued exchanging letters after Wallace left teaching to pursue his surveying career. They corresponded on prominent works of the time such as Charles Darwin's *The Voyage of the Beagle* (1839) and Robert Chamber's *Vestiges of the Natural History of Creation* (1844).

Wallace was inspired by the travelling naturalists of the day and decided to begin his exploration career collecting specimens in the Amazon rainforest. He explored the Rio Negra for four years, making notes on the peoples and languages he encountered as well as the geography, flora, and fauna. On his return voyage his ship, Helen, caught fire and he and the crew were stranded for ten days before being picked up by the Jordeson, a brig travelling from Cuba to London. All of his specimens aboard Helen had been lost.

After a brief stay in England he embarked on a journey to the Malay Archipelago (now Singapore, Malaysia, and Indonesia). During this eight year period he collected more than 126,000 specimens, several thousand of which represented new species to science. While travelling, Wallace refined his thoughts about evolution and in 1858 he outlined his theory of natural selection in an article he sent to Charles Darwin. This was published in the same year along with Darwin's own theory. Wallace eventually published an account of his travels *The Malay Archipelago* in 1869, and it became one of the most popular books of scientific exploration in the 19th century.

Upon his return to England, in 1862, Wallace became a staunch defender of Darwin's landmark work *On the Origin of Species* (1859). He wrote responses to those critical of the theory of natural selection, including 'Remarks on the Rev. S. Haughton's Paper on the Bee's Cell, And on the Origin of Species' (1863) and 'Creation by Law' (1867). The former of these was particularly pleasing to Darwin. Wallace also published important papers such as 'The Origin of Human Races and the Antiquity of Man Deduced from the Theory of 'Natural Selection" (1864) and books, including the much cited *Darwinism* (1889).

Wallace made a huge contribution to the natural sciences and he will continue to be remembered as one of the key figures in the development of evolutionary theory.

Wallace died on 7th November 1913 at the age of 90. He is buried in a small cemetery at Broadstone, Dorset, England.

CONTENTS

PREFACE. 9

I. PHYSICAL GEOGRAPHY. 17

II. SINGAPORE. 45

III. MALACCA AND MOUNT OPHIR 53

IV. BORNEO—THE ORANGUTAN. 66

V. BORNEO—JOURNEY INTO THE INTERIOR. 109

VI. BORNEO—THE DYAKS. 140

VII. JAVA. 151

VIII. SUMATRA. 189

IX. NATURAL HISTORY OF THE INDO-MALAY ISLANDS. 209

X. BALI AND LOMBOCK. 227

XI. LOMBOCK: MANNERS AND CUSTOMS OF THE PEOPLE. 248

XII. LOMBOCK: HOW THE RAJAH TOOK THE CENSUS. 268

XIII. TIMOR. 279

XIV. THE NATURAL HISTORY OF THE TIMOR GROUP. 305

XV. CELEBES. 320

XVI. CELEBES. 345

XVII. CELEBES. 363

XVIII. NATURAL HISTORY OF CELEBES. 406

XIX. BANDA. 428

XX. AMBOYNA. 438

PREFACE.

My readers will naturally ask why I have delayed writing this book for six years after my return; and I feel bound to give them full satisfaction on this point.

When I reached England in the spring of 1862, I found myself surrounded by a room full of packing cases containing the collections that I had, from time to time, sent home for my private use. These comprised nearly three thousand bird-skins of about one thousand species, at least twenty thousand beetles and butterflies of about seven thousand species, and some quadrupeds and land shells besides. A large proportion of these I had not seen for years, and in my then weakened state of health, the unpacking, sorting, and arranging of such a mass of specimens occupied a long time.

I very soon decided that until I had done something towards naming and describing the most important groups in my collection, and had worked out some of the more interesting problems of variation and geographical distribution (of which I had had glimpses while collecting them), I would not attempt to publish my travels. Indeed, I could have printed my notes and journals at once, leaving all reference to questions of natural history for a future work; but, I felt that this would be as unsatisfactory to myself as it would be disappointing to my friends, and uninstructive to the public.

Since my return, up to this date, I have published eighteen papers in the "Transactions" or "Proceedings of the Linnean Zoological and Entomological Societies", describing or cataloguing portions of my collections, along with twelve others in various scientific periodicals on more general subjects connected with them.

Nearly two thousand of my Coleoptera, and many hundreds of my butterflies, have been already described by various eminent naturalists, British and foreign; but a much larger number remains undescribed. Among those to whom science is most indebted for this laborious work, I must name Mr. F. P. Pascoe, late President of the Entomological Society of London, who had almost completed the classification and description of my large collection of Longicorn beetles (now in his possession), comprising more than a thousand species, of which at least nine hundred were previously undescribed and new to European cabinets.

The remaining orders of insects, comprising probably more than two thousand species, are in the collection of Mr. William Wilson Saunders, who has caused the larger portion of them to be described by good entomologists. The Hymenoptera alone amounted to more than nine hundred species, among which were two hundred and eighty different kinds of ants, of which two hundred were new.

Volume I

The six years' delay in publishing my travels thus enables me to give what I hope may be an interesting and instructive sketch of the main results yet arrived at by the study of my collections; and as the countries I have to describe are not much visited or written about, and their social and physical conditions are not liable to rapid change, I believe and hope that my readers will gain much more than they will lose by not having read my book six years ago, and by this time perhaps forgotten all about it.

I must now say a few words on the plan of my work.

My journeys to the various islands were regulated by the seasons and the means of conveyance. I visited some islands two or three times at distant intervals, and in some cases had to make the same voyage four times over. A chronological arrangement would have puzzled my readers. They would never have known where they were, and my frequent references to the groups of islands, classed in accordance with the peculiarities of their animal productions and of their human inhabitants, would have been hardly intelligible. I have adopted, therefore, a geographical, zoological, and ethnological arrangement, passing from island to island in what seems the most natural succession, while I transgress the order in which I myself visited them, as little as possible.

I divide the Archipelago into five groups of islands, as follows:

I. THE INDO-MALAY ISLANDS: comprising the Malay Peninsula and Singapore, Borneo, Java, and Sumatra.

II. THE TIMOR GROUP: comprising the islands of Timor, Flores, Sumbawa, and Lombock, with several smaller ones.

III. CELEBES: comprising also the Sula Islands and Bouton.

IV. THE MOLUCCAN GROUP: comprising Bouru, Ceram, Batchian, Gilolo, and Morty; with the smaller islands of Ternate, Tidore, Makian, Kaióa, Amboyna, Banda, Goram, and Matabello.

V. THE PAPUAN GROUP: comprising the great island of New Guinea, with the Aru Islands, Mysol, Salwatty, Waigiou, and several others. The Ke Islands are described with this group on account of their ethnology, though zoologically and geographically they belong to the Moluccas.

The chapters relating to the separate islands of each of these groups are followed by one on the Natural History of that group; and the work may thus be divided into five parts, each treating one of the natural divisions of the Archipelago.

Volume I

The first chapter is an introductory one, on the Physical Geography of the whole region; and the last is a general sketch of the races of man in the Archipelago and the surrounding countries. With this explanation, and a reference to the maps which illustrate the work, I trust that my readers will always know where they are, and in what direction they are going.

I am well aware that my book is far too small for the extent of the subjects it touches upon. It is a mere sketch; but so far as it goes, I have endeavoured to make it an accurate one. Almost the whole of the narrative and descriptive portions were written on the spot, and have had little more than verbal alterations. The chapters on Natural History, as well as many passages in other parts of the work, have been written in the hope of exciting an interest in the various questions connected with the origin of species and their geographical distribution. In some cases I have been able to explain my views in detail; while in others, owing to the greater complexity of the subject, I have thought it better to confine myself to a statement of the more interesting facts of the problem, whose solution is to be found in the principles developed by Mr. Darwin in his various works. The numerous illustrations will, it is believed, add much to the interest and value of the book. They have been made from my own sketches, from photographs, or from specimens—and such, only subjects that would really illustrate the narrative or the descriptions, have been chosen.

I have to thank Messrs. Walter and Henry Woodbury, whose acquaintance I had the pleasure of making in Java, for a number of photographs of scenery and of natives, which have been of the greatest assistance to me. Mr. William Wilson Saunders has kindly allowed me to figure the curious horned flies; and to Mr. Pascoe I am indebted for a loan of two of the very rare Longicorns which appear in the plate of Bornean beetles. All the other specimens figured are in my own collection.

As the main object of all my journeys was to obtain specimens of natural history, both for my private collection and to supply duplicates to museums and amateurs, I will give a general statement of the number of specimens I collected, and which reached home in good condition. I must premise that I generally employed one or two, and sometimes three Malay servants to assist me; and for nearly half the time had the services of an English lad, Charles Allen. I was just eight years away from England, but as I travelled about fourteen thousand miles within the Archipelago, and made sixty or seventy separate journeys, each involving some preparation and loss of time, I do not think that more than six years were really occupied in collecting.

I find that my Eastern collections amounted to:

310 specimens of Mammalia.
100 specimens of Reptiles.

Volume I

8,050 specimens of Birds.
7,500 specimens of Shells.
13,100 specimens of Lepidoptera.
83,200 specimens of Coleoptera.
13,400 specimens of other Insects.

125,660 specimens of natural history in all.

It now only remains for me to thank all those friends to whom I am indebted for assistance or information. My thanks are more especially due to the Council of the Royal Geographical Society, through whose valuable recommendations I obtained important aid from our own Government and from that of Holland; and to Mr. William Wilson Saunders, whose kind and liberal encouragement in the early portion of my journey was of great service to me. I am also greatly indebted to Mr. Samuel Stevens (who acted as my agent), both for the care he took of my collections, and for the untiring assiduity with which he kept me supplied, both with useful information and with whatever necessaries I required.

I trust that these, and all other friends who have been in any way interested in my travels and collections, may derive from the perusal of my book, some faint reflexion of the pleasures I myself enjoyed amid the scenes and objects it describes.

THE MALAY ARCHIPELAGO.

CHAPTER I. PHYSICAL GEOGRAPHY.

From a look at a globe or a map of the Eastern hemisphere, we shall perceive between Asia and Australia a number of large and small islands forming a connected group distinct from those great masses of land, and having little connection with either of them. Situated upon the Equator, and bathed by the tepid water of the great tropical oceans, this region enjoys a climate more uniformly hot and moist than almost any other part of the globe, and teems with natural productions which are elsewhere unknown. The richest of fruits and the most precious of spices are Indigenous here. It produces the giant flowers of the Rafflesia, the great green-winged Ornithoptera (princes among the butterfly tribes), the man-like Orangutan, and the gorgeous Birds of Paradise. It is inhabited by a peculiar and interesting race of mankind—the Malay, found nowhere beyond the limits of this insular tract, which has hence been named the Malay Archipelago.

To the ordinary Englishman this is perhaps the least

known part of the globe. Our possessions in it are few and scanty; scarcely any of our travellers go to explore it; and in many collections of maps it is almost ignored, being divided between Asia and the Pacific Islands. It thus happens that few persons realize that, as a whole, it is comparable with the primary divisions of the globe, and that some of its separate islands are larger than France or the Austrian Empire. The traveller, however, soon acquires different ideas. He sails for days or even weeks along the shores of one of these great islands, often so great that its inhabitants believe it to be a vast continent. He finds that voyages among these islands are commonly reckoned by weeks and months, and that their several inhabitants are often as little known to each other as are the native races of the northern to those of the southern continent of America. He soon comes to look upon this region as one apart from the rest of the world, with its own races of men and its own aspects of nature; with its own ideas, feelings, customs, and modes of speech, and with a climate, vegetation, and animated life altogether peculiar to itself.

From many points of view these islands form one compact geographical whole, and as such they have always been treated by travellers and men of science; but, a more careful and detailed study of them under various aspects reveals the unexpected fact that they are divisible into two portions nearly equal in extent which differ widely in their natural products, and really form two parts of the primary divisions of the earth. I have been able to prove this in considerable detail by my observations on the natural history of the various parts of the Archipelago; and, as in the description of my travels and residence in the several islands I shall have to refer continually to this view, and adduce facts in support of it, I have thought it advisable to commence with a general sketch of the main features of the Malayan region as will render the facts hereafter brought forward more interesting, and their bearing upon the general question more easily understood. I proceed, therefore, to sketch the limits and extent of the Archipelago, and to point out the more striking features of its geology, physical geography, vegetation, and animal life.

Definition and Boundaries.—For reasons which depend mainly on the distribution of animal life, I consider the Malay Archipelago to include the Malay Peninsula as far as Tenasserim and the Nicobar Islands on the west, the Philippines on the north, and the Solomon Islands, beyond New Guinea, on the east. All the great islands included within these limits are connected together by innumerable smaller ones, so that no one of them seems to be distinctly separated from the rest. With but few exceptions all enjoy an uniform and very similar climate, and are covered with a luxuriant forest vegetation. Whether we study their form and distribution on maps, or actually travel from island to island, our first impression will be that they form a connected whole, all the parts of which are intimately related to each other.

Extent of the Archipelago and Islands.—The Malay Archipelago extends for more than 4,000 miles in length from east to west, and is about 1,300 in breadth from north to south. It would stretch over an expanse equal to that of all Europe from the extreme west far into Central Asia, or would cover the widest parts of South America, and extend far beyond the land into the Pacific and Atlantic oceans. It includes three islands larger than Great Britain; and in one of them, Borneo, the whole of the British Isles might be set down, and would be surrounded by a sea of forests. New Guinea, though less compact in shape, is probably larger than Borneo. Sumatra is about equal in extent to Great Britain; Java, Luzon, and Celebes are each about the size of Ireland. Eighteen more islands are, on the average, as large as Jamaica; more than a hundred are as large as the Isle of Wight; while the isles and islets of smaller size are innumerable.

The absolute extent of land in the Archipelago is not greater than that contained by Western Europe from Hungary to Spain; but, owing to the manner in which the land is broken up and divided, the variety of its productions is rather in proportion to the immense surface over which the islands are spread, than to the quantity of land which they contain.

Volume I

Geological Contrasts.—One of the chief volcanic belts upon the globe passes through the Archipelago, and produces a striking contrast in the scenery of the volcanic and non-volcanic islands. A curving line, marked out by scores of active, and hundreds of extinct, volcanoes may be traced through the whole length of Sumatra and Java, and thence by the islands of Bali, Lombock, Sumbawa, Flores, the Serwatty Islands, Banda, Amboyna, Batchian, Makian, Tidore, Ternate, and Gilolo, to Morty Island. Here there is a slight but well-marked break, or shift, of about 200 miles to the westward, where the volcanic belt begins again in North Celebes, and passes by Siau and Sanguir to the Philippine Islands along the eastern side of which it continues, in a curving line, to their northern extremity. From the extreme eastern bend of this belt at Banda, we pass onwards for 1,000 miles over a non-volcanic district to the volcanoes observed by Dampier, in 1699, on the north-eastern coast of New Guinea, and can there trace another volcanic belt through New Britain, New Ireland, and the Solomon Islands, to the eastern limits of the Archipelago.

In the whole region occupied by this vast line of volcanoes, and for a considerable breadth on each side of it, earthquakes are of continual recurrence, slight shocks being felt at intervals of every few weeks or months, while more severe ones, shaking down whole villages, and doing more or less injury to life and property, are sure to happen, in one part or another of this district, almost every year. On many of the islands the years of the great earthquakes form the chronological epochs of the native inhabitants, by the aid of which the ages of their children are remembered, and the dates of many important events are determined.

    I can only briefly allude to the many fearful eruptions that have taken place in this region. In the amount of injury to life and property, and in the magnitude of their effects, they have not been surpassed by any upon record. Forty villages were destroyed by the eruption of Papandayang in Java, in 1772, when the whole mountain was blown up by repeated explosions, and a large lake left in its place. By the great eruption of Tomboro in Sumbawa, in 1815, 12,000 people were destroyed, and the ashes darkened the air and fell thickly upon the earth and sea for 300 miles around. Even quite recently, since I left the country, a mountain which had been quiescent for more than 200 years suddenly burst into activity. The island of Makian, one of the Moluccas, was rent open in 1646 by a violent eruption which left a huge chasm on one side, extending into the heart of the mountain. It was, when I last visited it in 1860, clothed with vegetation to the summit, and contained twelve populous Malay villages. On the 29th of December, 1862, after 215 years of perfect inaction, it again suddenly burst forth, blowing up and completely altering the appearance of the mountain, destroying the greater part of the inhabitants, and sending forth such volumes of ashes as to darken the air at Ternate, forty miles off, and to almost entirely destroy the growing crops on that and the surrounding islands.

The island of Java contains more volcanoes, active and extinct, than any other known district of equal extent. They are about forty-five in number, and many of them exhibit most beautiful examples of the volcanic cone on a large scale, single or double, with entire or truncated summits, and averaging 10,000 feet high.

It is now well ascertained that almost all volcanoes have been slowly built up by the accumulation of matter—mud, ashes, and lava—ejected by themselves. The openings or craters, however, frequently shift their position, so that a country may be covered with a more or less irregular series of hills in chains and masses, only here and there rising into lofty cones, and yet the whole may be produced by true volcanic action. In this manner the greater part of Java has been formed. There has been some elevation, especially on the south coast, where extensive cliffs of coral limestone are found; and there may be a substratum of older stratified rocks;¹ but still essentially Java is volcanic, and that noble and fertile island—the very garden of the East, and perhaps upon the whole the richest, the best cultivated, and the best governed tropical island in the world—owes its very existence to the same intense volcanic activity which still occasionally devastates its surface.

The great island of Sumatra exhibits, in proportion to its extent, a much smaller number of volcanoes, and a considerable portion of it has probably a non-volcanic origin.

To the eastward, the long string of islands from Java, passing by the north of Timor and away to Banda, are probably all due to volcanic action. Timor itself consists of ancient stratified rocks, but is said to have one volcano near its centre.

Going northward, Amboyna, a part of Bouru, and the west end of Ceram, the north part of Gilolo, and all the small islands around it, the northern extremity of Celebes, and the islands of Siau and Sanguir, are wholly volcanic. The Philippine Archipelago contains many active and extinct volcanoes, and has probably been reduced to its present fragmentary condition by subsidences attending on volcanic action.

All along this great line of volcanoes are to be found more or less palpable signs of upheaval and depression of land. The range of islands south of Sumatra, a part of the south coast of Java and of the islands east of it, the west and east end of Timor, portions of all the Moluccas, the Ke and Aru Islands, Waigiou, and the whole south and east of Gilolo, consist in a great measure of upraised coral-rock, exactly corresponding to that now forming in the adjacent seas. In many places I have observed the unaltered surfaces of the elevated reefs, with great masses of coral standing up in their natural position, and hundreds of shells so fresh-looking that it was hard to believe that they had been more than a few years out of the water; and, in fact, it is very probable that such changes have occurred within a few centuries.

The united lengths of these volcanic belts is about ninety degrees, or one-fourth of the entire circumference of the globe. Their width is about fifty miles; but, for a space of two hundred miles on each side of them, evidences of subterranean action are to be found in recently elevated coral-rock, or in barrier coral-reefs, indicating recent submergence. In the very centre or focus of the great curve of volcanoes is placed the large island of Borneo, in which no sign of recent volcanic action has yet been observed, and where earthquakes, so characteristic of the surrounding regions, are entirely unknown. The equally large island of New Guinea occupies another quiescent area, on which no sign of volcanic action has yet been discovered. With the exception of the eastern end of its northern peninsula, the large and curiously-shaped island of Celebes is also entirely free from volcanoes; and there is some reason to believe that the volcanic portion has once formed a separate island. The Malay Peninsula is also non-volcanic.

The first and most obvious division of the Archipelago would therefore be into quiescent and volcanic regions, and it might, perhaps, be expected that such a division would correspond to some differences in the character of the vegetation and the forms of life. This is the case, however, to a very limited extent; and we shall presently see that, although this development of subterranean fires is on so vast a scale—has piled up chains of mountains ten or twelve thousand feet high—has broken up continents and raised up islands from the ocean—yet it has all the character of a recent action which has not yet succeeded in obliterating the traces of a more ancient distribution of land and water.

Contrasts of Vegetation.—Placed immediately upon the Equator and surrounded by extensive oceans, it is not surprising that the various islands of the Archipelago should be almost always clothed with a forest vegetation from the level of the sea to the summits of the loftiest mountains. This is the general rule. Sumatra, New Guinea, Borneo, the Philippines and the Moluccas, and the uncultivated parts of Java and Celebes, are all forest countries, except a few small and unimportant tracts, due perhaps, in some cases, to ancient cultivation or accidental fires. To this, however, there is one important exception in the island of Timor and all the smaller islands around it, in which there is absolutely no forest such as exists in the other islands, and this character extends in a lesser degree to Flores, Sumbawa, Lombock, and Bali.

In Timor the most common trees are Eucalypti of several

species, also characteristic of Australia, with sandalwood, acacia, and other sorts in less abundance. These are scattered over the country more or less thickly, but, never so as to deserve the name of a forest. Coarse and scanty grasses grow beneath them on the more barren hills, and a luxuriant herbage in the moister localities. In the islands between Timor and Java there is often a more thickly wooded country abounding in thorny and prickly trees. These seldom reach any great height, and during the force of the dry season they almost completely lose their leaves, allowing the ground beneath them to be parched up, and contrasting strongly with the damp, gloomy, ever-verdant forests of the other islands. This peculiar character, which extends in a less degree to the southern peninsula of Celebes and the east end of Java, is most probably owing to the proximity of Australia. The south-east monsoon, which lasts for about two-thirds of the year (from March to November), blowing over the northern parts of that country, produces a degree of heat and dryness which assimilates the vegetation and physical aspect of the adjacent islands to its own. A little further eastward in Timor and the Ke Islands, a moister climate prevails; the southeast winds blowing from the Pacific through Torres Straits and over the damp forests of New Guinea, and as a consequence, every rocky islet is clothed with verdure to its very summit. Further west again, as the same dry winds blow over a wider and wider extent of ocean, they have time to absorb fresh moisture, and we accordingly find the island of Java possessing a less and less arid climate, until in the extreme west near Batavia, rain occurs more or less all the year round, and the mountains are everywhere clothed with forests of unexampled luxuriance.

Contrasts in Depth of Sea.—It was first pointed out by Mr. George Windsor Earl, in a paper read before the Royal Geographical Society in 1845, and subsequently in a pamphlet "On the Physical Geography of South-Eastern Asia and Australia", dated 1855, that a shallow sea connected the great islands of Sumatra, Java, and Borneo with the Asiatic continent, with which their natural productions generally agreed; while a similar shallow sea connected New Guinea and some of the adjacent islands to Australia, all being characterised by the presence of marsupials.

We have here a clue to the most radical contrast in the Archipelago, and by following it out in detail I have arrived at the conclusion that we can draw a line among the islands, which shall so divide them that one-half shall truly belong to Asia, while the other shall no less certainly be allied to Australia. I term these respectively the Indo-Malayan and the Austro-Malayan divisions of the Archipelago.

On referring to pages 12, 13, and 36 of Mr. Earl's pamphlet, it will be seen that he maintains the former connection of Asia and Australia as an important part of his view; whereas, I dwell mainly on their long continued separation. Notwithstanding this and other important differences between us, to him undoubtedly belongs the merit of first indicating the division of the Archipelago into an Australian and an Asiatic region, which it has been my good fortune to establish by more detailed observations.

Contrasts in Natural Productions.—To understand the importance of this class of facts, and its bearing upon the former distribution of land and sea, it is necessary to consider the results arrived at by geologists and naturalists in other parts of the world.

It is now generally admitted that the present distribution of living things on the surface of the earth is mainly the result of the last series of changes that it has undergone. Geology teaches us that the surface of the land, and the distribution of land and water, is everywhere slowly changing. It further teaches us that the forms of life which inhabit that surface have, during every period of which we possess any record, been also slowly changing.

It is not now necessary to say anything about how either of those changes took place; as to that, opinions may differ; but as to the fact that the changes themselves have occurred, from the earliest geological ages down to the present day, and are still going on, there is no difference of opinion. Every successive stratum of sedimentary rock, sand, or gravel, is a proof that changes of level have taken place; and the different species of animals and plants, whose remains are found in these deposits, prove that corresponding changes did occur in the organic world.

Taking, therefore, these two series of changes for granted, most of the present peculiarities and anomalies in the distribution of species may be directly traced to them. In our own islands, with a very few trifling exceptions, every quadruped, bird, reptile, insect, and plant, is found also on the adjacent continent. In the small islands of Sardinia and Corsica, there are some quadrupeds and insects, and many plants, quite peculiar. In Ceylon, more closely connected to India than Britain is to Europe, many animals and plants are different from those found in India, and peculiar to the island. In the Galapagos Islands, almost every indigenous living thing is peculiar to them, though closely resembling other kinds found in the nearest parts of the American continent.

Most naturalists now admit that these facts can only be explained by the greater or less lapse of time since the islands were upraised from beneath the ocean, or were separated from the nearest land; and this will be generally (though not always) indicated by the depth of the intervening sea. The enormous thickness of many marine deposits through wide areas shows that subsidence has often continued (with intermitting periods of repose) during epochs of immense duration. The depth of sea produced by such subsidence will therefore generally be a measure of time; and in like manner, the change which organic forms have undergone is a measure of time. When we make proper allowance for the continued introduction of new animals and plants from surrounding countries by those natural means of dispersal which have been so well explained by Sir Charles Lyell and Mr. Darwin, it is remarkable how closely these two measures correspond. Britain is separated from the continent by a very shallow sea, and only in a very few cases have our animals or plants begun to show a difference from the corresponding continental species. Corsica and Sardinia, divided from Italy by a much deeper sea, present a much greater difference in their organic forms. Cuba, separated from Yucatan by a wider and deeper strait, differs more markedly, so that most of its productions are of distinct and peculiar species; while Madagascar, divided from Africa by a deep channel three hundred miles wide, possesses so many peculiar features as to indicate separation at a very remote antiquity, or even to render it doubtful whether the two countries have ever been absolutely united.

Returning now to the Malay Archipelago, we find that all the wide expanse of sea which divides Java, Sumatra, and Borneo from each other, and from Malacca and Siam, is so shallow that ships can anchor in any part of it, since it rarely exceeds forty fathoms in depth; and if we go as far as the line of a hundred fathoms, we shall include the Philippine Islands and Bali, east of Java. If, therefore, these islands have been separated from each other and the continent by subsidence of the intervening tracts of land, we should conclude that the separation has been comparatively recent, since the depth to which the land has subsided is so small. It is also to be remarked that the great chain of active volcanoes in Sumatra and Java furnishes us with a sufficient cause for such subsidence, since the enormous masses of matter they have thrown out would take away the foundations of the surrounding district; and this may be the true explanation of the often-noticed fact that volcanoes and volcanic chains are always near the sea. The subsidence they produce around them will, in time, make a sea, if one does not already exist.

But, it is when we examine the zoology of these countries that we find what we most require—evidence of a very striking character that these great islands must have once formed a part of the continent, and could only have been separated at a very recent geological epoch. The elephant and tapir of Sumatra and Borneo, the rhinoceros of Sumatra and the allied species of Java, the wild cattle of Borneo and the kind long supposed to be peculiar to Java, are now all known to inhabit some part or other of Southern Asia. None of these large animals could possibly have passed over the arms of the sea which now separate these countries, and their presence plainly indicates that a land communication must have existed since the origin of the species. Among the smaller mammals, a considerable portion are common to each island and the continent; but the vast physical changes that must have occurred during the breaking up and subsidence of such extensive regions have led to the extinction of some in one or more of the islands, and in some cases there seems also to have been time for a change of species to have taken place. Birds and insects illustrate the same view, for every family and almost every genus of these groups found in any of the islands occurs also on the Asiatic continent, and in a great number of cases the species are exactly identical. Birds offer us one of the best means of determining the law of distribution; for though at first sight it would appear that the watery boundaries which keep out the land quadrupeds could be easily passed over by birds, yet practically it is not so; for if we leave out the aquatic tribes which are pre-eminently wanderers, it is found that the others (and especially the Passeres, or true perching-birds, which form the vast majority) are generally as strictly limited by straits and arms of the sea as are quadrupeds themselves. As an instance, among the islands of which I am now speaking, it is a remarkable fact that Java possesses numerous birds which never pass over to Sumatra, though they are separated by a strait only fifteen miles wide, and with islands in mid-channel. Java, in fact, possesses more birds and insects peculiar to itself than either Sumatra or Borneo, and this would indicate that it was earliest separated from the continent; next in organic individuality is Borneo, while Sumatra is so nearly identical in all its animal forms with the peninsula of Malacca, that we may safely conclude it to have been the most recently dismembered island.

The general result therefore, at which we arrive, is that the great islands of Java, Sumatra, and Borneo resemble in their natural productions the adjacent parts of the continent, almost as much as such widely-separated districts could be expected to do even if they still formed a part of Asia; and this close resemblance, joined with the fact of the wide extent of sea which separates them being so uniformly and remarkably shallow, and lastly, the existence of the extensive range of volcanoes in Sumatra and Java, which have poured out vast quantities of subterranean matter and have built up extensive plateaux and lofty mountain ranges, thus furnishing a vera causa for a parallel line of subsidence—all lead irresistibly to the conclusion that at a very recent geological epoch, the continent of Asia extended far beyond its present limits in a south-easterly direction, including the islands of Java, Sumatra, and Borneo, and probably reaching as far as the present 100-fathom line of soundings.

The Philippine Islands agree in many respects with Asia and the other islands, but present some anomalies, which seem to indicate that they were separated at an earlier period, and have since been subject to many revolutions in their physical geography.

Turning our attention now to the remaining portion of

the Archipelago, we shall find that all the islands from Celebes and Lombock eastward exhibit almost as close a resemblance to Australia and New Guinea as the Western Islands do to Asia. It is well known that the natural productions of Australia differ from those of Asia more than those of any of the four ancient quarters of the world differ from each other. Australia, in fact, stands alone: it possesses no apes or monkeys, no cats or tigers, wolves, bears, or hyenas; no deer or antelopes, sheep or oxen; no elephant, horse, squirrel, or rabbit; none, in short, of those familiar types of quadruped which are met with in every other part of the world. Instead of these, it has Marsupials only: kangaroos and opossums; wombats and the duckbilled Platypus. In birds it is almost as peculiar. It has no woodpeckers and no pheasants—families which exist in every other part of the world; but instead of them it has the mound-making brush-turkeys, the honeysuckers, the cockatoos, and the brush-tongued lories, which are found nowhere else upon the globe. All these striking peculiarities are found also in those islands which form the Austro-Malayan division of the Archipelago.

The great contrast between the two divisions of the Archipelago is nowhere so abruptly exhibited as on passing from the island of Bali to that of Lombock, where the two regions are in closest proximity. In Bali we have barbets, fruit-thrushes, and woodpeckers; on passing over to Lombock these are seen no more, but we have abundance of cockatoos, honeysuckers, and brush-turkeys, which are equally unknown in Bali, or any island further west. [I was informed, however, that there were a few cockatoos at one spot on the west of Bali, showing that the intermingling of the productions of these islands is now going on.] The strait is here fifteen miles wide, so that we may pass in two hours from one great division of the earth to another, differing as essentially in their animal life as Europe does from America. If we travel from Java or Borneo to Celebes or the Moluccas, the difference is still more striking. In the first, the forests abound in monkeys of many kinds, wild cats, deer, civets, and otters, and numerous varieties of squirrels are constantly met with. In the latter none of these occur; but the prehensile-tailed Cuscus is almost the only terrestrial mammal seen, except wild pigs, which are found in all the islands, and deer (which have probably been recently introduced) in Celebes and the Moluccas. The birds which are most abundant in the Western Islands are woodpeckers, barbets, trogons, fruit-thrushes, and leaf-thrushes; they are seen daily, and form the great ornithological features of the country. In the Eastern Islands these are absolutely unknown, honeysuckers and small lories being the most common birds, so that the naturalist feels himself in a new world, and can hardly realize that he has passed from the one region to the other in a few days, without ever being out of sight of land.

The inference that we must draw from these facts is,

undoubtedly, that the whole of the islands eastwards beyond Java and Borneo do essentially form a part of a former Australian or Pacific continent, although some of them may never have been actually joined to it. This continent must have been broken up not only before the Western Islands were separated from Asia, but probably before the extreme southeastern portion of Asia was raised above the waters of the ocean; for a great part of the land of Borneo and Java is known to be geologically of quite recent formation, while the very great difference of species, and in many cases of genera also, between the productions of the Eastern Malay Islands and Australia, as well as the great depth of the sea now separating them, all point to a comparatively long period of isolation.

It is interesting to observe among the islands themselves how a shallow sea always intimates a recent land connexion. The Aru Islands, Mysol, and Waigiou, as well as Jobie, agree with New Guinea in their species of mammalia and birds much more closely than they do with the Moluccas, and we find that they are all united to New Guinea by a shallow sea. In fact, the 100-fathom line round New Guinea marks out accurately the range of the true Paradise birds.

It is further to be noted—and this is a very interesting point in connection with theories of the dependence of special forms of life on external conditions—that this division of the Archipelago into two regions characterised by a striking diversity in their natural productions does not in any way correspond to the main physical or climatal divisions of the surface. The great volcanic chain runs through both parts, and appears to produce no effect in assimilating their productions. Borneo closely resembles New Guinea not only in its vast size and its freedom from volcanoes, but in its variety of geological structure, its uniformity of climate, and the general aspect of the forest vegetation that clothes its surface. The Moluccas are the counterpart of the Philippines in their volcanic structure, their extreme fertility, their luxuriant forests, and their frequent earthquakes; and Bali with the east end of Java has a climate almost as dry and a soil almost as arid as that of Timor. Yet between these corresponding groups of islands, constructed as it were after the same pattern, subjected to the same climate, and bathed by the same oceans, there exists the greatest possible contrast when we compare their animal productions. Nowhere does the ancient doctrine—that differences or similarities in the various forms of life that inhabit different countries are due to corresponding physical differences or similarities in the countries themselves—meet with so direct and palpable a contradiction. Borneo and New Guinea, as alike physically as two distinct countries can be, are zoologically wide as the poles asunder; while Australia, with its dry winds, its open plains, its stony deserts, and its temperate climate, yet produces birds and quadrupeds which are closely related to those inhabiting the hot damp luxuriant forests, which everywhere clothe the plains and mountains of New Guinea.

In order to illustrate more clearly the means by which I suppose this great contrast has been brought about, let us consider what would occur if two strongly contrasted divisions of the earth were, by natural means, brought into proximity. No two parts of the world differ so radically in their productions as Asia and Australia, but the difference between Africa and South America is also very great, and these two regions will well serve to illustrate the question we are considering. On the one side we have baboons, lions, elephants, buffaloes, and giraffes; on the other spider-monkeys, pumas, tapirs, anteaters, and sloths; while among birds, the hornbills, turacos, orioles, and honeysuckers of Africa contrast strongly with the toucans, macaws, chatterers, and hummingbirds of America.

Now let us endeavour to imagine (what it is very probable may occur in future ages) that a slow upheaval of the bed of the Atlantic should take place, while at the same time earthquake-shocks and volcanic action on the land should cause increased volumes of sediment to be poured down by the rivers, so that the two continents should gradually spread out by the addition of newly-formed lands, and thus reduce the Atlantic which now separates them, to an arm of the sea a few hundred miles wide. At the same time we may suppose islands to be upheaved in mid-channel; and, as the subterranean forces varied in intensity, and shifted their points of greatest action, these islands would sometimes become connected with the land on one side or other of the strait, and at other times again be separated from it. Several islands would at one time be joined together, at another would be broken up again, until at last, after many long ages of such intermittent action, we might have an irregular archipelago of islands filling up the ocean channel of the Atlantic, in whose appearance and arrangement we could discover nothing to tell us which had been connected with Africa and which with America. The animals and plants inhabiting these islands would, however, certainly reveal this portion of their former history. On those islands which had ever formed a part of the South American continent, we should be sure to find such common birds as chatterers and toucans and hummingbirds, and some of the peculiar American quadrupeds; while on those which had been separated from Africa, hornbills, orioles, and honeysuckers would as certainly be found. Some portion of the upraised land might at different times have had a temporary connection with both continents, and would then contain a certain amount of mixture in its living inhabitants. Such seems to have been the case with the islands of Celebes and the Philippines. Other islands, again, though in such close proximity as Bali and Lombock, might each exhibit an almost unmixed sample of the productions of the continents of which they had directly or indirectly once formed a part.

In the Malay Archipelago we have, I believe, a case exactly parallel to that which I have here supposed. We have indications of a vast continent, with a peculiar fauna and flora having been gradually and irregularly broken up; the island of Celebes probably marking its furthest westward extension, beyond which was a wide ocean. At the same time Asia appears to have been extending its limits in a southeast direction, first in an unbroken mass, then separated into islands as we now see it, and almost coming into actual contact with the scattered fragments of the great southern land.

From this outline of the subject, it will be evident how important an adjunct Natural History is to Geology; not only in interpreting the fragments of extinct animals found in the earth's crust, but in determining past changes in the surface which have left no geological record. It is certainly a wonderful and unexpected fact that an accurate knowledge of the distribution of birds and insects should enable us to map out lands and continents which disappeared beneath the ocean long before the earliest traditions of the human race. Wherever the geologist can explore the earth's surface, he can read much of its past history, and can determine approximately its latest movements above and below the sea-level; but wherever oceans and seas now extend, he can do nothing but speculate on the very limited data afforded by the depth of the waters. Here the naturalist steps in, and enables him to fill up this great gap in the past history of the earth.

One of the chief objects of my travels was to obtain evidence of this nature; and my search after such evidence has been rewarded by great success, so that I have been able to trace out with some probability the past changes which one of the most interesting parts of the earth has undergone. It may be thought that the facts and generalizations here given would have been more appropriately placed at the end rather than at the beginning of a narrative of the travels which supplied the facts. In some cases this might be so, but I have found it impossible to give such an account as I desire of the natural history of the numerous islands and groups of islands in the Archipelago, without constant reference to these generalizations which add so much to their interest. Having given this general sketch of the subject, I shall be able to show how the same principles can be applied to the individual islands of a group, as to the whole Archipelago; and thereby make my account of the many new and curious animals which inhabit them both, more interesting and more instructive than if treated as mere isolated facts.

Contrasts of Races.—Before I had arrived at the conviction that the eastern and western halves of the Archipelago belonged to distinct primary regions of the earth, I had been led to group the natives of the Archipelago under two radically distinct races. In this I differed from most ethnologists who had before written on the subject; for it had been the almost universal custom to follow William von Humboldt and Pritchard, in classing all the Oceanic races as modifications of one type. Observation soon showed me, however, that Malays and Papuans differed radically in every physical, mental, and moral character; and more detailed research, continued for eight years, satisfied me that under these two forms, as types, the whole of the peoples of the Malay Archipelago and Polynesia could be classified. On drawing the line which separates these races, it is found to come near to that which divides the zoological regions, but somewhat eastward of it; a circumstance which appears to me very significant of the same causes having influenced the distribution of mankind that have determined the range of other animal forms.

The reason why exactly the same line does not limit both is sufficiently intelligible. Man has means of traversing the sea which animals do not possess; and a superior race has power to press out or assimilate an inferior one. The maritime enterprise and higher civilization of the Malay races have enabled them to overrun a portion of the adjacent region, in which they have entirely supplanted the indigenous inhabitants if it ever possessed any; and to spread much of their language, their domestic animals, and their customs far over the Pacific, into islands where they have but slightly, or not at all, modified the physical or moral characteristics of the people.

I believe, therefore, that all the peoples of the various islands can be grouped either with the Malays or the Papuans; and that these two have no traceable affinity to each other. I believe, further, that all the races east of the line I have drawn have more affinity for each other than they have for any of the races west of that line; that, in fact, the Asiatic races include the Malays, and all have a continental origin, while the Pacific races, including all to the east of the former (except perhaps some in the Northern Pacific), are derived, not from any existing continent, but from lands which now exist or have recently existed in the Pacific Ocean. These preliminary observations will enable the reader better to apprehend the importance I attach to the details of physical form or moral character, which I shall give in describing the inhabitants of many of the islands.

CHAPTER II. SINGAPORE.

(A SKETCH OF THE TOWN AND ISLAND AS SEEN DURING SEVERAL VISITS FROM 1854 TO 1862.)

FEW places are more interesting to a traveller from Europe than the town and island of Singapore, furnishing, as it does, examples of a variety of Eastern races, and of many different religions and modes of life. The government, the garrison, and the chief merchants are English; but the great mass of the population is Chinese, including some of the wealthiest merchants, the agriculturists of the interior, and most of the mechanics and labourers. The native Malays are usually fishermen and boatmen, and they form the main body of the police. The Portuguese of Malacca supply a large number of the clerks and smaller merchants. The Klings of Western India are a numerous body of Mahometans, and, with many Arabs, are petty merchants and shopkeepers. The grooms and washermen are all Bengalees, and there is a small but highly respectable class of Parsee merchants. Besides these, there are numbers of Javanese sailors and domestic servants, as well as traders from Celebes, Bali, and many other islands of the Archipelago. The harbour is crowded with men-of-war and trading vessels of many European nations, and hundreds of Malay praus and Chinese junks, from vessels of several hundred tons burthen down to little fishing boats and passenger sampans; and the town comprises handsome public buildings and churches, Mahometan mosques, Hindu temples, Chinese joss-houses, good European houses, massive warehouses, queer old Kling and China bazaars, and long suburbs of Chinese and Malay cottages.

By far the most conspicuous of the various kinds of people in Singapore, and those which most attract the stranger's attention, are the Chinese, whose numbers and incessant activity give the place very much the appearance of a town in China. The Chinese merchant is generally a fat round-faced man with an important and business-like look. He wears the same style of clothing (loose white smock, and blue or black trousers) as the meanest coolie, but of finer materials, and is always clean and neat; and his long tail tipped with red silk hangs down to his heels. He has a handsome warehouse or shop in town and a good house in the country. He keeps a fine horse and gig, and every evening may be seen taking a drive bareheaded to enjoy the cool breeze. He is rich—he owns several retail shops and trading schooners, he lends money at high interest and on good security, he makes hard bargains, and gets fatter and richer every year.

In the Chinese bazaar are hundreds of small shops in

which a miscellaneous collection of hardware and dry goods are to be found, and where many things are sold wonderfully cheap. You may buy gimlets at a penny each, white cotton thread at four balls for a halfpenny, and penknives, corkscrews, gunpowder, writing-paper, and many other articles as cheap or cheaper than you can purchase them in England. The shopkeeper is very good-natured; he will show you everything he has, and does not seem to mind if you buy nothing. He bates a little, but not so much as the Klings, who almost always ask twice what they are willing to take. If you buy a few things from him, he will speak to you afterwards every time you pass his shop, asking you to walk in and sit down, or take a cup of tea; and you wonder how he can get a living where so many sell the same trifling articles.

The tailors sit at a table, not on one; and both they and the shoemakers work well and cheaply. The barbers have plenty to do, shaving heads and cleaning ears; for which latter operation they have a great array of little tweezers, picks, and brushes. In the outskirts of the town are scores of carpenters and blacksmiths. The former seem chiefly to make coffins and highly painted and decorated clothes-boxes. The latter are mostly gun-makers, and bore the barrels of guns by hand out of solid bars of iron. At this tedious operation they may be seen every day, and they manage to finish off a gun with a flintlock very handsomely. All about the streets are sellers of water, vegetables, fruit, soup, and agar-agar (a jelly made of seaweed), who have many cries as unintelligible as those of London. Others carry a portable cooking-apparatus on a pole balanced by a table at the other end, and serve up a meal of shellfish, rice, and vegetables for two or three halfpence—while coolies and boatmen waiting to be hired are everywhere to be met with.

In the interior of the island the Chinese cut down forest trees in the jungle, and saw them up into planks; they cultivate vegetables, which they bring to market; and they grow pepper and gambir, which form important articles of export. The French Jesuits have established missions among these inland Chinese, which seem very successful. I lived for several weeks at a time with the missionary at Bukit-tima, about the centre of the island, where a pretty church has been built and there are about 300 converts. While there, I met a missionary who had just arrived from Tonquin, where he had been living for many years. The Jesuits still do their work thoroughly as of old. In Cochin China, Tonquin, and China, where all Christian teachers are obliged to live in secret, and are liable to persecution, expulsion, and sometimes death, every province—even those farthest in the interior—has a permanent Jesuit mission establishment constantly kept up by fresh aspirants, who are taught the languages of the countries they are going to at Penang or Singapore. In China there are said to be near a million converts; in Tonquin and

Cochin China, more than half a million. One secret of the success of these missions is the rigid economy practised in the expenditure of the funds. A missionary is allowed about £30. a year, on which he lives in whatever country he may be. This renders it possible to support a large number of missionaries with very limited means; and the natives, seeing their teachers living in poverty and with none of the luxuries of life, are convinced that they are sincere in what they teach, and have really given up home and friends and ease and safety, for the good of others. No wonder they make converts, for it must be a great blessing to the poor people among whom they labour to have a man among them to whom they can go in any trouble or distress, who will comfort and advise them, who visits them in sickness, who relieves them in want, and who they see living from day-to-day in danger of persecution and death—entirely for their sakes.

My friend at Bukit-tima was truly a father to his flock. He preached to them in Chinese every Sunday, and had evenings for discussion and conversation on religion during the week. He had a school to teach their children. His house was open to them day and night. If a man came to him and said, "I have no rice for my family to eat today," he would give him half of what he had in the house, however little that might be. If another said, "I have no money to pay my debt," he would give him half the contents of his purse, were it his last dollar. So, when he was himself in want, he would send to some of the wealthiest among his flock, and say, "I have no rice in the house," or "I have given away my money, and am in want of such and such articles." The result was that his flock trusted and loved him, for they felt sure that he was their true friend, and had no ulterior designs in living among them.

The island of Singapore consists of a multitude of small hills, three or four hundred feet high, the summits of many of which are still covered with virgin forest. The mission-house at Bukit-tima was surrounded by several of these wood-topped hills, which were much frequented by woodcutters and sawyers, and offered me an excellent collecting ground for insects. Here and there, too, were tiger pits, carefully covered over with sticks and leaves, and so well concealed, that in several cases I had a narrow escape from falling into them. They are shaped like an iron furnace, wider at the bottom than the top, and are perhaps fifteen or twenty feet deep so that it would be almost impossible for a person unassisted to get out of one. Formerly a sharp stake was stuck erect in the bottom; but after an unfortunate traveller had been killed by falling on one, its use was forbidden. There are always a few tigers roaming about Singapore, and they kill on an average a Chinaman every day, principally those who work in the gambir plantations, which are always made in newly-cleared jungle. We heard a tiger roar once or twice in the evening, and it was rather nervous work hunting for insects among the fallen trunks and old sawpits when one of these savage animals might be lurking close by, awaiting an opportunity to spring upon us.

Several hours in the middle of every fine day were spent in these patches of forest, which were delightfully cool and shady by contrast with the bare open country we had to walk over to reach them. The vegetation was most luxuriant, comprising enormous forest trees, as well as a variety of ferns, caladiums, and other undergrowth, and abundance of climbing rattan palms. Insects were exceedingly abundant and very interesting, and every day furnished scores of new and curious forms.

In about two months I obtained no less than 700 species of beetles, a large proportion of which were quite new, and among them were 130 distinct kinds of the elegant Longicorns (Cerambycidae), so much esteemed by collectors. Almost all these were collected in one patch of jungle, not more than a square mile in extent, and in all my subsequent travels in the East I rarely if ever met with so productive a spot. This exceeding productiveness was due in part no doubt to some favourable conditions in the soil, climate, and vegetation, and to the season being very bright and sunny, with sufficient showers to keep everything fresh. But it was also in a great measure dependent, I feel sure, on the labours of the Chinese wood-cutters. They had been at work here for several years, and during all that time had furnished a continual supply of dry and dead and decaying leaves and bark, together with abundance of wood and sawdust, for the nourishment of insects and their larvae. This had led to the assemblage of a great variety of species in a limited space, and I was the first naturalist who had come to reap the harvest they had prepared. In the same place, and during my walks in other directions, I obtained a fair collection of butterflies and of other orders of insects, so that on the whole I was quite satisfied with these—my first attempts to gain a knowledge of the Natural History of the Malay Archipelago.

CHAPTER III. MALACCA AND MOUNT OPHIR

(JULY TO SEPTEMBER, 1854.)

BIRDS and most other kinds of animals being scarce at Singapore, I left it in July for Malacca, where I spent more than two months in the interior, and made an excursion to Mount Ophir. The old and picturesque town of Malacca is crowded along the banks of the small river, and consists of narrow streets of shops and dwelling houses, occupied by the descendants of the Portuguese, and by Chinamen. In the suburbs are the houses of the English officials and of a few Portuguese merchants, embedded in groves of palms and fruit-trees, whose varied and beautiful foliage furnishes a pleasing relief to the eye, as well as most grateful shade.

The old fort, the large Government House, and the ruins of a cathedral attest the former wealth and importance of this place, which was once as much the centre of Eastern trade as Singapore is now. The following description of it by Linschott, who wrote two hundred and seventy years ago, strikingly exhibits the change it has undergone:

"Malacca is inhabited by the Portuguese and by natives of the country, called Malays. The Portuguese have here a fortress, as at Mozambique, and there is no fortress in all the Indies, after those of Mozambique and Ormuz, where the captains perform their duty better than in this one. This place is the market of all India, of China, of the Moluccas, and of other islands around about—from all which places, as well as from Banda, Java, Sumatra, Siam, Pegu, Bengal, Coromandel, and India—arrive ships which come and go incessantly, charged with an infinity of merchandises. There would be in this place a much greater number of Portuguese if it were not for the inconvenience, and unhealthiness of the air, which is hurtful not only to strangers, but also to natives of the country. Thence it is that all who live in the country pay tribute of their health, suffering from a certain disease, which makes them lose either their skin or their hair. And those who escape consider it a miracle, which occasions many to leave the country, while the ardent desire of gain induces others to risk their health, and endeavour to endure such an atmosphere. The origin of this town, as the natives say, was very small, only having at the beginning, by reason of the unhealthiness of the air, but six or seven fishermen who inhabited it. But the number was increased by the meeting of fishermen from Siam, Pegu, and Bengal, who came and built a city, and established a peculiar language, drawn from the most elegant modes of speaking of other nations, so that in fact the language of the Malays is at present the most refined, exact, and celebrated of all the East. The name of

Malacca was given to this town, which, by the convenience of its situation, in a short time grew to such wealth, that it does not yield to the most powerful towns and regions around about. The natives, both men and women, are very courteous and are reckoned the most skillful in the world in compliments, and study much to compose and repeat verses and love-songs. Their language is in vogue through the Indies, as the French is here."

At present, a vessel over a hundred tons hardly ever enters its port, and the trade is entirely confined to a few petty products of the forests, and to the fruit, which the trees, planted by the old Portuguese, now produce for the enjoyment of the inhabitants of Singapore. Although rather subject to fevers, it is not at present considered very unhealthy.

The population of Malacca consists of several races. The ubiquitous Chinese are perhaps the most numerous, keeping up their manners, customs, and language; the indigenous Malays are next in point of numbers, and their language is the Lingua-franca of the place. Next come the descendants of the Portuguese—a mixed, degraded, and degenerate race, but who still keep up the use of their mother tongue, though ruefully mutilated in grammar; and then there are the English rulers, and the descendants of the Dutch, who all speak English. The Portuguese spoken at Malacca is a useful philological phenomenon. The verbs have mostly lost their inflections, and one form does for all moods, tenses, numbers, and persons. Eu vai, serves for "I go," "I went," or, "I will go." Adjectives, too, have been deprived of their feminine and plural terminations, so that the language is reduced to a marvellous simplicity, and, with the admixture of a few Malay words, becomes rather puzzling to one who has heard only the pure Lusitanian.

In costume these several peoples are as varied as in their speech. The English preserve the tight-fitting coat, waistcoat, and trousers, and the abominable hat and cravat; the Portuguese patronise a light jacket, or, more frequently, shirt and trousers only; the Malays wear their national jacket and sarong (a kind of kilt), with loose drawers; while the Chinese never depart in the least from their national dress, which, indeed, it is impossible to improve for a tropical climate, whether as regards comfort or appearance. The loosely-hanging trousers, and neat white half-shirt half-jacket, are exactly what a dress should be in this low latitude.

I engaged two Portuguese to accompany me into the interior; one as a cook, the other to shoot and skin birds, which is quite a trade in Malacca. I first stayed a fortnight at a village called Gading, where I was accommodated in the house of some Chinese converts, to whom I was recommended by the Jesuit missionaries. The house was a mere shed, but it was kept clean, and I made myself sufficiently comfortable. My hosts were forming a pepper and gambir plantation, and in the immediate neighbourhood were extensive tin-washings, employing over a thousand Chinese. The tin is obtained in the form of black grains from beds of quartzose sand, and is melted into ingots in rude clay furnaces. The soil seemed poor, and the forest was very dense with undergrowth, and not at all productive of insects; but, on the other hand, birds were abundant, and I was at once introduced to the rich ornithological treasures of the Malayan region.

The very first time I fired my gun I brought down one of the most curious and beautiful of the Malacca birds, the blue-billed gaper (Cymbirhynchus macrorhynchus), called by the Malays the "Rainbird." It is about the size of a starling, black and rich claret colour with white shoulder stripes, and a very large and broad bill of the most pure cobalt blue above and orange below, while the iris is emerald green. As the skins dry the bill turns dull black, but even then the bird is handsome. When fresh killed, the contrast of the vivid blue with the rich colours of the plumage is remarkably striking and beautiful. The lovely Eastern trogons, with their rich-brown backs, beautifully pencilled wings, and crimson breasts, were also soon obtained, as well as the large green barbets (Megalaema versicolor)—fruit-eating birds, something like small toucans, with a short, straight bristly bill, and whose head and neck are variegated with patches of the most vivid blue and crimson. A day or two after, my hunter brought me a specimen of the green gaper (Calyptomena viridis), which is like a small cock-of-the-rock, but entirely of the most vivid green, delicately marked on the wings with black bars. Handsome woodpeckers and gay kingfishers, green and brown cuckoos with velvety red faces and green beaks, red-breasted doves and metallic honeysuckers, were brought in day after day, and kept me in a continual state of pleasurable excitement. After a fortnight one of my servants was seized with fever, and on returning to Malacca, the same disease, attacked the other as well as myself. By a liberal use of quinine, I soon recovered, and obtaining other men, went to stay at the Government bungalow of Ayer-panas, accompanied by a young gentleman, a native of the place, who had a taste for natural history.

At Ayer-panas we had a comfortable house to stay in, and plenty of room to dry and preserve our specimens; but, owing to there being no industrious Chinese to cut down timber, insects were comparatively scarce, with the exception of butterflies, of which I formed a very fine collection. The manner in which I obtained one fine insect was curious, and indicates how fragmentary and imperfect a traveller's collection must necessarily be. I was one afternoon walking along a favourite road through the forest, with my gun, when I saw a butterfly on the ground. It was large, handsome, and quite new to me, and I got close to it before it flew away. I then observed that it had been settling on the dung of some carnivorous animal. Thinking it might return to the same spot, I next day after breakfast took my net, and as I approached the place was delighted to see the same butterfly sitting on the same piece of dung, and succeeded in capturing it. It was an entirely new species of great beauty, and has been named by Mr. Hewitson—Nymphalis calydona. I never saw another specimen of it, and it was only after twelve years had elapsed that a second individual reached this country from the northwestern part of Borneo.

Having determined to visit Mount Ophir, which is situated in the middle of the peninsula about fifty miles east of Malacca, we engaged six Malays to accompany us and carry our baggage. As we meant to stay at least a week at the mountain, we took with us a good supply of rice, a little biscuit, butter and coffee, some dried fish and a little brandy, with blankets, a change of clothes, insect and bird boxes, nets, guns and ammunition. The distance from Ayer-panas was supposed to be about thirty miles.

Our first day's march lay through patches of forest, clearings, and Malay villages, and was pleasant enough. At night we slept at the house of a Malay chief, who lent us a verandah, and gave us a fowl and some eggs. The next day the country got wilder and more hilly. We passed through extensive forests, along paths often up to our knees in mud, and were much annoyed by the leeches for which this district is famous. These little creatures infest the leaves and herbage by the side of the paths, and when a passenger comes along they stretch themselves out at full length, and if they touch any part of his dress or body, quit their leaf and adhere to it. They then creep on to his feet, legs, or other part of his body and suck their fill, the first puncture being rarely felt during the excitement of walking. On bathing in the evening we generally found half a dozen or a dozen on each of us, most frequently on our legs, but sometimes on our bodies, and I had one who sucked his fill from the side of my neck, but who luckily missed the jugular vein. There are many species of these forest leeches. All are small, but some are beautifully marked with stripes of bright yellow. They probably attach themselves to deer or other animals which frequent the forest paths, and have thus acquired the singular habit of stretching themselves out at the sound of a footstep or of rustling foliage. Early in the afternoon we reached the foot of the mountain, and encamped by the side of a fine stream, whose rocky banks were overgrown with ferns. Our oldest Malay had been accustomed to shoot birds in this neighbourhood for the Malacca dealers, and had been to the top of the mountain, and while we amused ourselves shooting and insect hunting, he went with two others to clear the path for our ascent the next day.

Early next morning we started after breakfast, carrying blankets and provisions, as we intended to sleep upon the mountain. After passing a little tangled jungle and swampy thickets through which our men had cleared a path, we emerged into a fine lofty forest pretty clear of undergrowth, and in which we could walk freely. We ascended steadily up a moderate slope for several miles, having a deep ravine on our left. We then had a level plateau or shoulder to cross, after which the ascent was steeper and the forest denser until we came out upon the "Padang-batu," or stone field, a place of which we had heard much, but could never get anyone to describe intelligibly. We found it to be a steep slope of even rock, extending along the mountain side farther than we could see. Parts of it were quite bare, but where it was cracked and fissured there grew a most luxuriant vegetation, among which the pitcher plants were the most remarkable. These wonderful plants never seem to succeed well in our hot-houses, and are there seen to little advantage. Here they grew up into half climbing shrubs, their curious pitchers of various sizes and forms hanging abundantly from their leaves, and continually exciting our admiration by their size and beauty. A few coniferae of the genus Dacrydium here first appeared, and in the thickets just above the rocky surface we walked through groves of those splendid ferns Dipteris Horsfieldii and Matonia pectinata, which bear large spreading palmate fronds on slender stems six or eight feet high. The Matonia is the tallest and most elegant, and is known only from this mountain, and neither of them is yet introduced into our hot-houses.

It was very striking to come out from the dark, cool,

and shady forest in which we had been ascending since we started, on to this hot, open rocky slope where we seemed to have entered at one step from a lowland to an alpine vegetation. The height, as measured by a sympiesometer, was about 2,800 feet. We had been told we should find water at Padang-batu as we were exceedingly thirsty; but we looked about for it in vain. At last we turned to the pitcher-plants, but the water contained in the pitchers (about half a pint in each) was full of insects, and otherwise uninviting. On tasting it, however, we found it very palatable though rather warm, and we all quenched our thirst from these natural jugs. Farther on we came to forest again, but of a more dwarf and stunted character than below; and alternately passing along ridges and descending into valleys, we reached a peak separated from the true summit of the mountain by a considerable chasm. Here our porters gave in, and declared they could carry their loads no further; and certainly the ascent to the highest peak was very precipitous. But on the spot where we were there was no water, whereas it was well known that there was a spring close to the summit, so we determined to go on without them, and carry with us only what was absolutely necessary. We accordingly took a blanket each, and divided our food and other articles among us, and went on with only the old Malay and his son.

After descending into the saddle between the two peaks we found the ascent very laborious, the slope being so steep, as often to necessitate hand-climbing. Besides a bushy vegetation the ground was covered knee-deep with mosses on a foundation of decaying leaves and rugged rock, and it was a hard hour's climb to the small ledge just below the summit, where an overhanging rock forms a convenient shelter, and a little basin collects the trickling water. Here we put down our loads, and in a few minutes more stood on the summit of Mount Ophir, 4,000 feet above the sea. The top is a small rocky platform covered with rhododendrons and other shrubs. The afternoon was clear, and the view fine in its way—ranges of hill and valley everywhere covered with interminable forest, with glistening rivers winding among them.

In a distant view a forest country is very monotonous, and no mountain I have ever ascended in the tropics presents a panorama equal to that from Snowdon, while the views in Switzerland are immeasurably superior. When boiling our coffee I took observations with a good boiling-point thermometer, as well as with the sympiesometer, and we then enjoyed our evening meal and the noble prospect that lay before us. The night was calm and very mild, and having made a bed of twigs and branches over which we laid our blankets, we passed a very comfortable night. Our porters had followed us after a rest, bringing only their rice to cook, and luckily we did not require the baggage they left behind them. In the morning I caught a few butterflies and beetles, and my friend got a few land-shells; and we then descended, bringing with us some specimens of the ferns and pitcher-plants of Padang-batu.

The place where we had first encamped at the foot of the mountain being very gloomy, we chose another in a kind of swamp near a stream overgrown with Zingiberaceous plants, in which a clearing was easily made. Here our men built two little huts without sides that would just shelter us from the rain; we lived in them for a week, shooting and insect-hunting, and roaming about the forests at the foot of the mountain. This was the country of the great Argus pheasant, and we continually heard its cry. On asking the old Malay to try and shoot one for me, he told me that although he had been for twenty years shooting birds in these forests he had never yet shot one, and had never even seen one except after it had been caught. The bird is so exceedingly shy and wary, and runs along the ground in the densest parts of the forest so quickly, that it is impossible to get near it; and its sober colours and rich eye-like spots, which are so ornamental when seen in a museum, must harmonize well with the dead leaves among which it dwells, and render it very inconspicuous. All the specimens sold in Malacca are caught in snares, and my informant, though he had shot none, had snared plenty.

The tiger and rhinoceros are still found here, and a few years ago elephants abounded, but they have lately all disappeared. We found some heaps of dung, which seemed to be that of elephants, and some tracks of the rhinoceros, but saw none of the animals. However, we kept a fire up all night in case any of these creatures should visit us, and two of our men declared that they did one day see a rhinoceros. When our rice was finished, and our boxes full of specimens, we returned to Ayer-Panas, and a few days afterwards went on to Malacca, and thence to Singapore. Mount Ophir has quite a reputation for fever, and all our friends were astonished at our recklessness in staying so long at its foot; but none of us suffered in the least, and I shall ever look back with pleasure to my trip as being my first introduction to mountain scenery in the Eastern tropics.

The meagreness and brevity of the sketch I have here given of my visit to Singapore and the Malay Peninsula is due to my having trusted chiefly to some private letters and a notebook, which were lost; and to a paper on Malacca and Mount Ophir which was sent to the Royal Geographical Society, but which was neither read nor printed owing to press of matter at the end of a session, and the MSS. of which cannot now be found. I the less regret this, however, as so many works have been written on these parts; and I always intended to pass lightly over my travels in the western and better known portions of the Archipelago, in order to devote more space to the remoter districts, about which hardly anything has been written in the English language.

CHAPTER IV. BORNEO—THE ORANGUTAN.

I ARRIVED at Sarawak on November 1st, 1854, and left it on January 25th, 1856. In the interval I resided at many different localities, and saw a good deal of the Dyak tribes as well as of the Bornean Malays. I was hospitably entertained by Sir James Brooke, and lived in his house whenever I was at the town of Sarawak in the intervals of my journeys. But so many books have been written about this part of Borneo since I was there, that I shall avoid going into details of what I saw and heard and thought of Sarawak and its ruler, confining myself chiefly to my experiences as a naturalist in search of shells, insects, birds and the Orangutan, and to an account of a journey through a part of the interior seldom visited by Europeans.

The first four months of my visit were spent in various parts of the Sarawak River, from Santubong at its mouth up to the picturesque limestone mountains and Chinese goldfields of Bow and Bede. This part of the country has been so frequently described that I shall pass it over, especially as, owing to its being the height of the wet season, my collections were comparatively poor and insignificant.

In March 1865 I determined to go to the coalworks

which were being opened near the Simunjon River, a small branch of the Sadong, a river east of Sarawak and between it and the Batang-Lupar. The Simunjon enters the Sadong River about twenty miles up. It is very narrow and very winding, and much overshadowed by the lofty forest, which sometimes almost meets over it. The whole country between it and the sea is a perfectly level forest-covered swamp, out of which rise a few isolated hills, at the foot of one of which the works are situated. From the landing-place to the hill a Dyak road had been formed, which consisted solely of tree-trunks laid end to end. Along these the barefooted natives walk and carry heavy burdens with the greatest ease, but to a booted European it is very slippery work, and when one's attention is constantly attracted by the various objects of interest around, a few tumbles into the bog are almost inevitable. During my first walk along this road I saw few insects or birds, but noticed some very handsome orchids in flower, of the genus Coelogyne, a group which I afterwards found to be very abundant, and characteristic of the district. On the slope of the hill near its foot a patch of forest had been cleared away, and several rude houses erected, in which were residing Mr. Coulson the engineer, and a number of Chinese workmen. I was at first kindly accommodated in Mr. Coulson's house, but finding the spot very suitable for me and offering great facilities for collecting, I had a small house of two rooms and a verandah built for myself. Here

I remained nearly nine months, and made an immense collection of insects, to which class of animals I devoted my chief attention, owing to the circumstances being especially favourable.

In the tropics a large proportion of the insects of all orders, and especially of the large and favourite group of beetles, are more or less dependent on vegetation, and particularly on timber, bark, and leaves in various stages of decay. In the untouched virgin forest, the insects which frequent such situations are scattered over an immense extent of country, at spots where trees have fallen through decay and old age, or have succumbed to the fury of the tempest; and twenty square miles of country may not contain so many fallen and decayed trees as are to be found in any small clearing. The quantity and the variety of beetles and of many other insects that can be collected at a given time in any tropical locality, will depend, first upon the immediate vicinity of a great extent of virgin forest, and secondly upon the quantity of trees that for some months past have been, and which are still being cut down, and left to dry and decay upon the ground.

Now, during my whole twelve years' collecting in the western and eastern tropics, I never enjoyed such advantages in this respect as at the Simunjon coalworks. For several months from twenty to fifty Chinamen and Dyaks were employed almost exclusively in clearing a large space in the forest, and in making a wide opening for a railroad to the Sadong River, two miles distant. Besides this, sawpits were established at various points in the jungle, and large trees were felled to be cut up into beams and planks. For hundreds of miles in every direction a magnificent forest extended over plain and mountain, rock and morass, and I arrived at the spot just as the rains began to diminish and the daily sunshine to increase; a time which I have always found the most favourable season for collecting. The number of openings, sunny places, and pathways were also an attraction to wasps and butterflies; and by paying a cent each for all insects that were brought me, I obtained from the Dyaks and the Chinamen many fine locusts and Phasmidae, as well as numbers of handsome beetles.

When I arrived at the mines, on the 14th of March, I had collected in the four preceding months, 320 different kinds of beetles. In less than a fortnight I had doubled this number, an average of about 24 new species every day. On one day I collected 76 different kinds, of which 34 were new to me. By the end of April I had more than a thousand species, and they then went on increasing at a slower rate, so that I obtained altogether in Borneo about two thousand distinct kinds, of which all but about a hundred were collected at this place, and on scarcely more than a square mile of ground. The most numerous and most interesting groups of beetles were the Longicorns and Rhynchophora, both pre-eminently wood-feeders. The former, characterised by their graceful forms and long antenna, were especially numerous, amounting to nearly three hundred species, nine-tenths of which were entirely new, and many of them remarkable for their large size, strange forms, and beautiful colouring. The latter correspond to our weevils and allied groups, and in the tropics are exceedingly numerous and varied, often swarming upon dead timber, so that I sometimes obtained fifty or sixty different kinds in a day. My Bornean collections of this group exceeded five hundred species.

My collection of butterflies was not large; but I obtained some rare and very handsome insects, the most remarkable being the Ornithoptera Brookeana, one of the most elegant species known. This beautiful creature has very long and pointed wings, almost resembling a sphinx moth in shape. It is deep velvety black, with a curved band of spots of a brilliant metallic-green colour extending across the wings from tip to tip, each spot being shaped exactly like a small triangular feather, and having very much the effect of a row of the wing coverts of the Mexican trogon, laid upon black velvet. The only other marks are a broad neck-collar of vivid crimson, and a few delicate white touches on the outer margins of the hind wings. This species, which was then quite new and which I named after Sir James Brooke, was very rare. It was seen occasionally flying swiftly in the clearings, and now and then settling for an instant at puddles and muddy places, so that I only succeeded in capturing two or three specimens. In some other parts of the country I was assured it was abundant, and a good many specimens have been sent to England; but as yet all have been males, and we are quite unable to conjecture what the female may be like, owing to the extreme isolation of the species, and its want of close affinity to any other known insect.

One of the most curious and interesting reptiles which I met with in Borneo was a large tree-frog, which was brought me by one of the Chinese workmen. He assured me that he had seen it come down in a slanting direction from a high tree, as if it flew. On examining it, I found the toes very long and fully webbed to their very extremity, so that when expanded they offered a surface much larger than the body. The forelegs were also bordered by a membrane, and the body was capable of considerable inflation. The back and limbs were of a very deep shining green colour, the undersurface and the inner toes yellow, while the webs were black, rayed with yellow. The body was about four inches long, while the webs of each hind foot, when fully expanded, covered a surface of four square inches, and the webs of all the feet together about twelve square inches. As the extremities of the toes have dilated discs for adhesion, showing the creature to be a true tree frog, it is difficult to imagine that this immense membrane of the toes can be for the purpose of swimming only, and the account of the Chinaman, that it flew down from the tree, becomes more credible. This is, I believe, the first instance known of a "flying frog," and it is very interesting to Darwinians as showing that the variability of the toes which have been already modified for purposes of swimming and adhesive climbing, have been taken advantage of to enable an allied species to pass through the air like the flying lizard. It would appear to be a new species of the genus Rhacophorus, which consists of several frogs of a much smaller size than this, and having the webs of the toes less developed.

During my stay in Borneo I had no hunter to shoot for me regularly, and, being myself fully occupied with insects, I did not succeed in obtaining a very good collection of the birds or Mammalia, many of which, however, are well known, being identical with species found in Malacca. Among the Mammalia were five squirrels, and two tigercats—the Gymnurus Rafflesii, which looks like a cross between a pig and a polecat, and the Cynogale Bennetti—a rare, otter-like animal, with very broad muzzle clothed with long bristles.

One of my chief objects in coming to stay at Simunjon was to see the Orangutan (or great man-like ape of Borneo) in his native haunts, to study his habits, and obtain good specimens of the different varieties and species of both sexes, and of the adult and young animals. In all these objects I succeeded beyond my expectations, and will now give some account of my experience in hunting the Orangutan, or

"Mias," as it is called by the natives; and as this name is short, and easily pronounced, I shall generally use it in preference to Simia satyrus, or Orangutan.

Just a week after my arrival at the mines, I first saw a Mias. I was out collecting insects, not more than a quarter of a mile from the house, when I heard a rustling in a tree near, and, looking up, saw a large red-haired animal moving slowly along, hanging from the branches by its arms. It passed on from tree to tree until it was lost in the jungle, which was so swampy that I could not follow it. This mode of progression was, however, very unusual, and is more characteristic of the Hylobates than of the Orang. I suppose there was some individual peculiarity in this animal, or the nature of the trees just in this place rendered it the most easy mode of progression.

About a fortnight afterwards I heard that one was feeding in a tree in the swamp just below the house, and, taking my gun, was fortunate enough to find it in the same place. As soon as I approached, it tried to conceal itself among the foliage; but, I got a shot at it, and the second barrel caused it to fall down almost dead, the two balls having entered the body. This was a male, about half-grown, being scarcely three feet high. On April 26th, I was out shooting with two Dyaks, when we found another about the same size. It fell at the first shot, but did not seem much hurt, and immediately climbed up the nearest tree, when I fired, and it again fell, with a broken arm and a wound in the body. The two Dyaks now ran up to it, and each seized hold of a hand, telling me to cut a pole, and they would secure it. But although one arm was broken and it was only a half-grown animal, it was too strong for these young savages, drawing them up towards its mouth notwithstanding all their efforts, so that they were again obliged to leave go, or they would have been seriously bitten. It now began climbing up the tree again; and, to avoid trouble, I shot it through the heart.

On May 2nd, I again found one on a very high tree, when I had only a small 80-bore gun with me. However, I fired at it, and on seeing me it began howling in a strange voice like a cough, and seemed in a great rage, breaking off branches with its hands and throwing them down, and then soon made off over the tree-tops. I did not care to follow it, as it was swampy, and in parts dangerous, and I might easily have lost myself in the eagerness of pursuit.

On the 12th of May I found another, which behaved in a very similar manner, howling and hooting with rage, and throwing down branches. I shot at it five times, and it remained dead on the top of the tree, supported in a fork in such a manner that it would evidently not fall. I therefore returned home, and luckily found some Dyaks, who came back with me, and climbed up the tree for the animal. This was the first full-grown specimen I had obtained; but it was a female, and not nearly so large or remarkable as the full-grown males. It was, however, 3 ft. 6 in. high, and its arms stretched out to a width of 6 ft. 6 in. I preserved the skin of this specimen in a cask of arrack, and prepared a perfect skeleton, which was afterwards purchased for the Derby Museum.

Only four days afterwards some Dyaks saw another Mias near the same place, and came to tell me. We found it to be a rather large one, very high up on a tall tree. At the second shot it fell rolling over, but almost immediately got up again and began to climb. At a third shot it fell dead. This was also a full-grown female, and while preparing to carry it home, we found a young one face downwards in the bog. This little creature was only about a foot long, and had evidently been hanging to its mother when she first fell. Luckily it did not appear to have been wounded, and after we had cleaned the mud out of its mouth it began to cry out, and seemed quite strong and active. While carrying it home it got its hands in my beard, and grasped so tightly that I had great difficulty in getting free, for the fingers are habitually bent inwards at the last joint so as to form complete hooks. At this time it had not a single tooth, but a few days afterwards it cut its two lower front teeth. Unfortunately, I had no milk to give it, as neither Malays, Chinese nor Dyaks ever use the article, and I in vain inquired for any female animal that could suckle my little infant. I was therefore obliged to give it rice-water from a bottle with a quill in the cork, which after a few trials it learned to suck very well. This was very meagre diet, and the little creature did not thrive well on it, although I added sugar and cocoa-nut milk occasionally, to make it more nourishing. When I put my finger in its mouth it sucked with great vigour, drawing in its cheeks with all its might in the vain effort to extract some milk, and only after persevering a long time would it give up in disgust, and set up a scream very like that of a baby in similar circumstances.

When handled or nursed, it was very quiet and contented, but when laid down by itself would invariably cry; and for the first few nights was very restless and noisy. I fitted up a little box for a cradle, with a soft mat for it to lie upon, which was changed and washed every day; and I soon found it necessary to wash the little Mias as well. After I had done so a few times, it came to like the operation, and as soon as it was dirty would begin crying and not leave off until I took it out and carried it to the spout, when it immediately became quiet, although it would wince a little at the first rush of the cold water and make ridiculously wry faces while the stream was running over its head. It enjoyed the wiping and rubbing dry amazingly, and when I brushed its hair seemed to be perfectly happy, lying quite still with its arms and legs stretched out while I thoroughly brushed the long hair of its back and arms. For the first few days it clung desperately with all four hands to whatever it could lay hold of, and I had to be careful to keep my beard out of its way, as its fingers clutched hold of hair more tenaciously than anything else, and it was impossible to free myself without assistance. When restless, it would struggle about with its hands up in the air trying to find something to take hold of, and, when it had got a bit of stick or rag in two or three of its hands, seemed quite happy. For want of something else, it would often seize its own feet, and after a time it would constantly cross its arms and grasp with each hand the long hair that grew just below the opposite shoulder. The great tenacity of its grasp soon diminished, and I was obliged to invent some means to give it exercise and strengthen its limbs. For this purpose I made a short ladder of three or four rounds, on which I put it to hang for a quarter of an hour at a time. At first it seemed much pleased, but it could not get all four hands in a comfortable position, and, after changing about several times, would leave hold of one hand after the other, and drop onto the floor. Sometimes when hanging only by two hands, it would loose one, and cross it to the opposite shoulder, grasping its own hair; and, as this seemed much more agreeable than the stick, it would then loose the other and tumble down, when it would cross both and lie on its back quite contentedly, never seeming to be hurt by its numerous tumbles. Finding it so fond of hair, I endeavoured to make an artificial mother, by wrapping up a piece of buffalo-skin into a bundle, and suspending it about a foot from the floor. At first this seemed to suit it admirably, as it could sprawl its legs about and always find some hair, which it grasped with the greatest tenacity. I was now in hopes that I had made the little orphan quite happy; and so it seemed for some time, until it began to remember its lost parent, and try to suck. It would pull itself up close to the skin, and try about everywhere for a likely place; but, as it only succeeded in getting mouthfuls of hair and wool, it would be greatly disgusted, and scream violently, and, after two or three attempts, let go altogether. One day it got some wool into its throat, and I thought it would have choked, but after much gasping it recovered, and I was obliged to take the imitation mother to pieces again, and give up this last attempt to exercise the little creature.

After the first week I found I could feed it better with a spoon, and give it a little more varied and more solid food. Well-soaked biscuit mixed with a little egg and sugar, and sometimes sweet potatoes, were readily eaten; and it was a never-failing amusement to observe the curious changes of countenance by which it would express its approval or dislike of what was given to it. The poor little thing would lick its lips, draw in its cheeks, and turn up its eyes with an expression of the most supreme satisfaction when it had a mouthful particularly to its taste. On the other hand, when its food was not sufficiently sweet or palatable, it would turn the mouthful about with its tongue for a moment as if trying to extract what flavour there was, and then push it all out between its lips. If the same food was continued, it would set up a scream and kick about violently, exactly like a baby in a passion.

After I had had the little Mias about three weeks, I fortunately obtained a young hare-lip monkey (Macacus cynomolgus), which, though small, was very active, and could feed itself. I placed it in the same box with the Mias, and they immediately became excellent friends, neither exhibiting the least fear of the other. The little monkey would sit upon the other's stomach, or even on its face, without the least regard to its feelings. While I was feeding the Mias, the monkey would sit by, picking up all that was spilt, and occasionally putting out its hands to intercept the spoon; and as soon as I had finished would pick off what was left sticking to the Mias' lips, and then pull open its mouth and see if any still remained inside; afterwards lying down on the poor creature's stomach as on a comfortable cushion. The little helpless Mias would submit to all these insults with the most exemplary patience, only too glad to have something warm near it, which it could clasp affectionately in its arms. It sometimes, however, had its revenge; for when the monkey wanted to go away, the Mias would hold on as long as it could by the loose skin of its back or head, or by its tail, and it was only after many vigorous jumps that the monkey could make his escape.

It was curious to observe the different actions of these two animals, which could not have differed much in age. The Mias, like a very young baby, lying on its back quite helpless, rolling lazily from side to side, stretching out all four hands into the air, wishing to grasp something, but hardly able to guide its fingers to any definite object; and when dissatisfied, opening wide its almost toothless mouth, and expressing its wants by a most infantine scream. The little monkey, on the other hand, in constant motion, running and jumping about wherever it pleased, examining everything around it, seizing hold of the smallest object with the greatest precision, balancing itself on the edge of the box or running up a post, and helping itself to anything eatable that came in its way. There could hardly be a greater contrast, and the baby Mias looked more baby-like by the comparison.

When I had had it about a month, it began to exhibit some signs of learning to run alone. When laid upon the floor it would push itself along by its legs, or roll itself over, and thus make an unwieldy progression. When lying in the box it would lift itself up to the edge into almost an erect position, and once or twice succeeded in tumbling out. When left dirty, or hungry, or otherwise neglected, it would scream violently until attended to, varied by a kind of coughing or pumping noise very similar to that which is made by the adult animal. If no one was in the house, or its cries were not attended to, it would be quiet after a little while, but the moment it heard a footstep would begin again harder than ever.

After five weeks it cut its two upper front teeth, but in all this time it had not grown the least bit, remaining both in size and weight the same as when I first procured it. This was no doubt owing to the want of milk or other equally nourishing food. Rice-water, rice, and biscuits were but a poor substitute, and the expressed milk of the cocoa-nut which I sometimes gave it did not quite agree with its stomach. To this I imputed an attack of diarrhoea from which the poor little creature suffered greatly, but a small dose of castor-oil operated well, and cured it. A week or two afterwards it was again taken ill, and this time more seriously. The symptoms were exactly those of intermittent fever, accompanied by watery swellings on the feet and head. It lost all appetite for its food, and, after lingering for a week a most pitiable object, died, after being in my possession nearly three months. I much regretted the loss of my little pet, which I had at one time looked forward to bringing up to years of maturity, and taking home to England. For several months it had afforded me daily amusement by its curious ways and the inimitably ludicrous expression of its little countenance. Its weight was three pounds nine ounces, its height fourteen inches, and the spread of its arms twenty-three inches. I preserved its skin and skeleton, and in doing so found that when it fell from the tree it must have broken an arm and a leg, which had, however, united so rapidly that I had only noticed the hard swellings on the limbs where the irregular junction of the bones had taken place.

Exactly a week after I had caught this interesting little animal, I succeeded in shooting a full-grown male Orangutan. I had just come home from an entomologising excursion when Charles [Charles Allen, an English lad of sixteen, accompanied me as an assistant] rushed in out of breath with running and excitement, and exclaimed, interrupted by gasps, "Get the gun, sir,—be quick,—such a large Mias!" "Where is it?" I asked, taking hold of my gun as I spoke, which happened luckily to have one barrel loaded with ball. "Close by, sir—on the path to the mines—he can't get away." Two Dyaks chanced to be in the house at the time, so I called them to accompany me, and started off, telling Charley to bring all the ammunition after me as soon as possible. The path from our clearing to the mines led along the side of the hill a little way up its slope, and parallel with it at the foot a wide opening had been made for a road, in which several Chinamen were working, so that the animal could not escape into the swampy forest below without descending to cross the road or ascending to get round the clearings. We walked cautiously along, not making the least noise, and listening attentively for any sound which might betray the presence of the Mias, stopping at intervals to gaze upwards. Charley soon joined us at the place where he had seen the creature, and having taken the ammunition and put a bullet in the other barrel, we dispersed a little, feeling sure that it must be somewhere near, as it had probably descended the hill, and would not be likely to return again.

After a short time I heard a very slight rustling sound overhead, but on gazing up could see nothing. I moved about in every direction to get a full view into every part of the tree under which I had been standing, when I again heard the same noise but louder, and saw the leaves shaking as if caused by the motion of some heavy animal which moved off to an adjoining tree. I immediately shouted for all of them to come up and try and get a view, so as to allow me to have a shot. This was not an easy matter, as the Mias had a knack of selecting places with dense foliage beneath. Very soon, however, one of the Dyaks called me and pointed upwards, and on looking I saw a great red hairy body and a huge black face gazing down from a great height, as if wanting to know what was making such a disturbance below. I instantly fired, and he made off at once, so that I could not then tell whether I had hit him.

He now moved very rapidly and very noiselessly for so large an animal, so I told the Dyaks to follow and keep him in sight while I loaded. The jungle was here full of large angular fragments of rock from the mountain above, and thick with hanging and twisted creepers. Running, climbing, and creeping among these, we came up with the creature on the top of a high tree near the road, where the Chinamen had discovered him, and were shouting their astonishment with open mouths: "Ya Ya, Tuan; Orangutan, Tuan." Seeing that he could not pass here without descending, he turned up again towards the hill, and I got two shots, and following quickly, had two more by the time he had again reached the path, but he was always more or less concealed by foliage, and protected by the large branch on which he was walking. Once while loading I had a splendid view of him, moving along a large limb of a tree in a semi-erect posture, and showing it to be an animal of the largest size. At the path he got on to one of the loftiest trees in the forest, and we could see one leg hanging down useless, having been broken by a ball. He now fixed himself in a fork, where he was hidden by thick foliage, and seemed disinclined to move. I was afraid he would remain and die in this position, and as it was nearly evening. I could not have got the tree cut down that day. I therefore fired again, and he then moved off, and going up the hill was obliged to get on to some lower trees, on the branches of one of which he fixed himself in such a position that he could not fall, and lay all in a heap as if dead, or dying.

I now wanted the Dyaks to go up and cut off the branch he was resting on, but they were afraid, saying he was not dead, and would come and attack them. We then shook the adjoining tree, pulled the hanging creepers, and did all we could to disturb him, but without effect, so I thought it best to send for two Chinamen with axes to cut down the tree. While the messenger was gone, however, one of the Dyaks took courage and climbed towards him, but the Mias did not wait for him to get near, moving off to another tree, where he got on to a dense mass of branches and creepers which almost completely hid him from our view. The tree was luckily a small one, so when the axes came we soon had it cut through; but it was so held up by jungle ropes and climbers to adjoining trees that it only fell into a sloping position. The Mias did not move, and I began to fear that after all we should not get him, as it was near evening, and half a dozen more trees would have to be cut down before the one he was on would fall. As a last resource we all began pulling at the creepers, which shook the tree very much, and, after a few minutes, when we had almost given up all hope, down he came with a crash and a thud like the fall of a giant. And he was a giant, his head and body being fully as large as a man's. He was of the kind called by the Dyaks "Mias Chappan," or "Mias Pappan," which has the skin of the face broadened out to a ridge or fold at each side. His outstretched arms measured seven feet three inches across, and his height, measuring fairly from the top of the head to the heel was four feet two inches. The body just below the arms was three feet two inches round, and was quite as long as a man's, the legs being exceedingly short in proportion.

On examination we found he had been dreadfully wounded. Both legs were broken, one hip-joint and the root of the spine completely shattered, and two bullets were found flattened in his neck and jaws. Yet he was still alive when he fell. The two Chinamen carried him home tied to a pole, and I was occupied with Charley the whole of the next day preparing the skin and boiling the bones to make a perfect skeleton, which are now preserved in the Museum at Derby.

About ten days after this, on June 4th, some Dyaks came to tell us that the day before a Mias had nearly killed one of their companions. A few miles down the river there is a Dyak house, and the inhabitants saw a large Orang feeding on the young shoots of a palm by the riverside. On being alarmed he retreated towards the jungle which was close by, and a number of the men, armed with spears and choppers, ran out to intercept him. The man who was in front tried to run his spear through the animal's body, but the Mias seized it in his hands, and in an instant got hold of the man's arm, which he seized in his mouth, making his teeth meet in the flesh above the elbow, which he tore and lacerated in a dreadful manner. Had not the others been close behind, the man would have been more seriously injured, if not killed, as he was quite powerless; but they soon destroyed the creature with their spears and choppers. The man remained ill for a long time, and never fully recovered the use of his arm.

They told me the dead Mias was still lying where it had

been killed, so I offered them a reward to bring it up to our landing-place immediately, which they promised to do. They did not come, however, until the next day, and then decomposition had commenced, and great patches of the hair came off, so that it was useless to skin it. This I regretted much, as it was a very fine full-grown male. I cut off the head and took it home to clean, while I got my men to make a closed fence about five feet high around the rest of the body, which would soon be devoured by maggots, small lizards, and ants, leaving me the skeleton. There was a great gash in his face, which had cut deep into the bone, but the skull was a very fine one, and the teeth were remarkably large and perfect.

On June 18th I had another great success, and obtained a fine adult male. A Chinaman told me he had seen him feeding by the side of the path to the river, and I found him at the same place as the first individual I had shot. He was feeding on an oval green fruit having a fine red arillus, like the mace which surrounds the nutmeg, and which alone he seemed to eat, biting off the thick outer rind and dropping it in a continual shower. I had found the same fruit in the stomach of some others which I had killed. Two shots caused this animal to loose his hold, but he hung for a considerable time by one hand, and then fell flat on his face and was half buried in the swamp. For several minutes he lay groaning and panting, while we stood close around, expecting every breath to be his last. Suddenly, however, by a violent effort he raised himself up, causing us all to step back a yard or two, when, standing nearly erect, he caught hold of a small tree, and began to ascend it. Another shot through the back caused him to fall down dead. A flattened bullet was found in his tongue, having entered the lower part of the abdomen and completely traversed the body, fracturing the first cervical vertebra. Yet it was after this fearful wound that he had risen, and begun climbing with considerable facility. This also was a full-grown male of almost exactly the same dimensions as the other two I had measured.

On June 21st I shot another adult female, which was eating fruit in a low tree, and was the only one which I ever killed by a single ball.

On June 24th I was called by a Chinaman to shoot a Mias, which, he said, was on a tree close by his house, at the coal-mines. Arriving at the place, we had some difficulty in finding the animal, as he had gone off into the jungle, which was very rocky and difficult to traverse. At last we found him up a very high tree, and could see that he was a male of the largest size. As soon as I had fired, he moved higher up the tree, and while he was doing so I fired again; and we then saw that one arm was broken. He had now reached the very highest part of an immense tree, and immediately began breaking off boughs all around, and laying them across and across to make a nest. It was very interesting to see how well he had chosen his place, and how rapidly he stretched out his unwounded arm in every direction, breaking off good-sized boughs with the greatest ease, and laying them back across each other, so that in a few minutes he had formed a compact mass of foliage, which entirely concealed him from our sight. He was evidently going to pass the night here, and would probably get away early the next morning, if not wounded too severely. I therefore fired again several times, in hopes of making him leave his nest; but, though I felt sure I had hit him, as at each shot he moved a little, he would not go away. At length he raised himself up, so that half his body was visible, and then gradually sank down, his head alone remaining on the edge of the nest. I now felt sure he was dead, and tried to persuade the Chinaman and his companion to cut down the tree; but it was a very large one, and they had been at work all day, and nothing would induce them to attempt it. The next morning, at daybreak, I came to the place, and found that the Mias was evidently dead, as his head was visible in exactly the same position as before. I now offered four Chinamen a day's wages each to cut the tree down at once, as a few hours of sunshine would cause decomposition on the surface of the skin; but, after looking at it and trying it, they determined that it was very big and very hard, and would not attempt it. Had I doubled my offer, they would probably have accepted it, as it would not have been more than two or three hours' work; and had I been on a short visit only, I would have done so; but as I was a resident, and intended remaining several months longer, it would not have answered to begin paying too exorbitantly, or I should have got nothing done in the future at a lower rate.

For some weeks after, a cloud of flies could be seen all day, hovering over the body of the dead Mias; but in about a month all was quiet, and the body was evidently drying up under the influence of a vertical sun alternating with tropical rains. Two or three months later two Malays, on the offer of a dollar, climbed the tree and let down the dried remains. The skin was almost entirely enclosing the skeleton, and inside were millions of the pupa-cases of flies and other insects, with thousands of two or three species of small necrophagous beetles. The skull had been much shattered by balls, but the skeleton was perfect, except one small wristbone, which had probably dropped out and been carried away by a lizard.

Three days after I had shot this one and lost it, Charles found three small Orangs feeding together. We had a long chase after them, and had a good opportunity of seeing how they make their way from tree to tree by always choosing those limbs whose branches are intermingled with those of some other tree, and then grasping several of the small twigs together before they venture to swing themselves across. Yet they do this so quickly and certainly, that they make way among the trees at the rate of full five or six miles an hour, as we had continually to run to keep up with them. One of these we shot and killed, but it remained high up in the fork of a tree; and, as young animals are of comparatively little interest, I did not have the tree cut down to get it.

At this time I had the misfortune to slip among some fallen trees, and hurt my ankle; and, not being careful enough at first, it became a severe inflamed ulcer, which would not heal, and kept me a prisoner in the house the whole of July and part of August. When I could get out again, I determined to take a trip up a branch of the Simunjon River to Semabang, where there was said to be a large Dyak house, a mountain with abundance of fruit, and plenty of Orangs and fine birds. As the river was very narrow, and I was obliged to go in a very small boat with little luggage, I only took with me a Chinese boy as a servant. I carried a cask of medicated arrack to put Mias skins in, and stores and ammunition for a fortnight. After a few miles, the stream became very narrow and winding, and the whole country on each side was flooded. On the banks were an abundance of monkeys—the common Macacus cynomolgus, a black Semnopithecus, and the extraordinary long-nosed monkey (Nasalis larvatus), which is as large as a three-year old child, has a very long tail, and a fleshy nose longer than that of the biggest-nosed man. The further we went on the narrower and more winding the stream became; fallen trees sometimes blocked up our passage, and sometimes tangled branches and creepers met completely across it, and had to be cut away before we could get on. It took us two days to reach Semabang, and we hardly saw a bit of dry land all the way. In the latter part of the journey I could touch the bushes on each side for miles; and we were often delayed by the screw-pines (Pandanus), which grow abundantly in the water, falling across the stream. In other places dense rafts of floating grass completely filled up the channel, making our journey a constant succession of difficulties.

Near the landing-place we found a fine house, 250 feet long, raised high above the ground on posts, with a wide verandah and still wider platform of bamboo in front of it. Almost all the people, however, were away on some excursion after edible birds'-nests or bees'-wax, and there only remained in the house two or three old men and women with a lot of children. The mountain or hill was close by, covered with a complete forest of fruit-trees, among which the Durian and Mangosteen were very abundant; but the fruit was not yet quite ripe, except a little here and there. I spent a week at this place, going out everyday in various directions about the mountain, accompanied by a Malay, who had stayed with me while the other boatmen returned. For three days we found no Orangs, but shot a deer and several monkeys. On the fourth day, however, we found a Mias feeding on a very lofty Durian tree, and succeeded in killing it, after eight shots. Unfortunately it remained in the tree, hanging by its hands, and we were obliged to leave it and return home, as it was several miles off. As I felt pretty sure it would fall during the night, I returned to the place early the next morning, and found it on the ground beneath the tree. To my astonishment and pleasure, it appeared to be a different kind from any I had yet seen; for although a full-grown male, by its fully developed teeth and very large canines, it had no sign of the lateral protuberance on the face, and was about one-tenth smaller in all its dimensions than the other adult males. The upper incisors, however, appeared to be broader than in the larger species, a character distinguishing the Simia morio of Professor Owen, which he had described from the cranium of a female specimen. As it was too far to carry the animal home, I set to work and skinned the body on the spot, leaving the head, hands, and feet attached, to be finished at home. This specimen is now in the British Museum.

At the end of a week, finding no more Orangs, I returned home; and, taking in a few fresh stores, and this time accompanied by Charles, went up another branch of the river, very similar in character, to a place called Menyille, where there were several small Dyak houses and one large one. Here the landing place was a bridge of rickety poles, over a considerable distance of water; and I thought it safer to leave my cask of arrack securely placed in the fork of a tree. To prevent the natives from drinking it, I let several of them see me put in a number of snakes and lizards; but I rather think this did not prevent them from tasting it. We were accommodated here in the verandah of the large house, in which were several great baskets of dried human heads, the trophies of past generations of head-hunters. Here also there was a little mountain covered with fruit-trees, and there were some magnificent Durian trees close by the house, the fruit of which was ripe; and as the Dyaks looked upon me as a benefactor in killing the Mias, which destroys a great deal of their fruit, they let us eat as much as we liked; we revelled in this emperor of fruits in its greatest perfection.

The very day after my arrival in this place, I was so fortunate as to shoot another adult male of the small Orang, the Mias-kassir of the Dyaks. It fell when dead, but caught in a fork of the tree and remained fixed. As I was very anxious to get it, I tried to persuade two young Dyaks who were with me to cut down the tree, which was tall, perfectly straight and smooth-barked, and without a branch for fifty or sixty feet. To my surprise, they said they would prefer climbing up it, but it would be a good deal of trouble, and, after a little talking together, they said they would try. They first went to a clump of bamboo that stood near, and cut down one of the largest stems. From this they chopped off a short piece, and splitting it, made a couple of stout pegs, about a foot long and sharp at one end. Then cutting a thick piece of wood for a mallet, they drove one of the pegs into the tree and hung their weight upon it. It held, and this seemed to satisfy them, for they immediately began making a quantity of pegs of the same kind, while I looked on with great interest, wondering how they could possibly ascend such a lofty tree by merely driving pegs in it, the failure of any one of which at a good height would certainly cause their death. When about two dozen pegs were made, one of them began cutting some very long and slender bamboo from another clump, and also prepared some cord from the bark of a small tree. They now drove in a peg very firmly at about three feet from the ground, and bringing one of the long bamboos, stood it upright close to the tree, and bound it firmly to the two first pegs, by means of the bark cord and small notches near the head of each peg. One of the Dyaks now stood on the first peg and drove in a third, about level with his face, to which he tied the bamboo in the same way, and then mounted another step, standing on one foot, and holding by the bamboo at the peg immediately above him, while he drove in the next one. In this manner he ascended about twenty feet; when the upright bamboo was becoming thin, another was handed up by his companion, and this was joined by tying both bamboos to three or four of the pegs. When this was also nearly ended, a third was added, and shortly after, the lowest branches of the tree were reached, along which the young Dyak scrambled, and soon sent the

Mias tumbling down headlong. I was exceedingly struck by the ingenuity of this mode of climbing, and the admirable manner in which the peculiar properties of the bamboo were made available. The ladder itself was perfectly safe, since if any one peg were loose or faulty, and gave way, the strain would be thrown on several others above and below it. I now understood the use of the line of bamboo pegs sticking in trees, which I had often seen, and wondered for what purpose they could have been put there. This animal was almost identical in size and appearance with the one I had obtained at Semabang, and was the only other male specimen of the Simia morio which I obtained. It is now in the Derby Museum.

I afterwards shot two adult females and two young ones of different ages, all of which I preserved. One of the females, with several young ones, was feeding on a Durian tree with unripe fruit; and as soon as she saw us she began breaking off branches and the great spiny fruits with every appearance of rage, causing such a shower of missiles as effectually kept us from approaching too near the tree. This habit of throwing down branches when irritated has been doubted, but I have, as here narrated, observed it myself on at least three separate occasions. It was however always the female Mias who behaved in this way, and it may be that the male, trusting more to his great strength and his powerful canine teeth, is not afraid of any other animal, and does not want to drive them away, while the parental instinct of the female leads her to adopt this mode of defending herself and her young ones.

In preparing the skins and skeletons of these animals, I was much troubled by the Dyak dogs, which, being always kept in a state of semi-starvation, are ravenous for animal food. I had a great iron pan, in which I boiled the bones to make skeletons, and at night I covered this over with boards, and put heavy stones upon it; but the dogs managed to remove these and carried away the greater part of one of my specimens. On another occasion they gnawed away a good deal of the upper leather of my strong boots, and even ate a piece of my mosquito-curtain, where some lamp-oil had been spilt over it some weeks before.

On our return down the stream, we had the fortune to fall in with a very old male Mias, feeding on some low trees growing in the water. The country was flooded for a long distance, but so full of trees and stumps that the laden boat could not be got in among them, and if it could have been we should only have frightened the Mias away. I therefore got into the water, which was nearly up to my waist, and waded on until I was near enough for a shot. The difficulty then was to load my gun again, for I was so deep in the water that I could not hold the gun sloping enough to pour the powder in. I therefore had to search for a shallow place, and after several shots under these trying circumstances, I was delighted to see the monstrous animal roll over into the water. I now towed him after me to the stream, but the Malays objected to having the animal put into the boat, and he was so heavy that I could not do it without their help. I looked about for a place to skin him, but not a bit of dry ground was to be seen, until at last I found a clump of two or three old trees and stumps, between which a few feet of soil had collected just above the water, which was just large enough for us to drag the animal upon it. I first measured him, and found him to be by far the largest I had yet seen, for, though the standing height was the same as the others (4 feet 2 inches), the outstretched arms were 7 feet 9 inches, which was six inches more than the previous one, and the immense broad face was 13 1/2 inches wide, whereas the widest I had hitherto seen was only 11 1/2 inches. The girth of the body was 3 feet 7 1/2 inches. I am inclined to believe, therefore, that the length and strength of the arms, and the width of the face continues increasing to a very great age, while the standing height, from the sole of the foot to the crown of the head, rarely if ever exceeds 4 feet 2 inches.

As this was the last Mias I shot, and the last time I saw an adult living animal, I will give a sketch of its general habits, and any other facts connected with it. The Orangutan is known to inhabit Sumatra and Borneo, and there is every reason to believe that it is confined to these two great islands, in the former of which, however, it seems to be much more rare. In Borneo it has a wide range, inhabiting many districts on the southwest, southeast, northeast, and northwest coasts, but appears to be chiefly confined to the low and swampy forests. It seems, at first sight, very inexplicable that the Mias should be quite unknown in the Sarawak valley, while it is abundant in Sambas, on the west, and Sadong, on the east. But when we know the habits and mode of life of the animal, we see a sufficient reason for this apparent anomaly in the physical features of the Sarawak district. In the Sadong, where I observed it, the Mias is only found when the country is low level and swampy, and at the same time covered with a lofty virgin forest. From these swamps rise many isolated mountains, on some of which the Dyaks have settled and covered with plantations of fruit trees. These are a great attraction to the Mias, which comes to feed on the unripe fruits, but always retires to the swamp at night. Where the country becomes slightly elevated, and the soil dry, the Mias is no longer to be found. For example, in all the lower part of the Sadong valley it abounds, but as soon as we ascend above the limits of the tides, where the country, though still flat, is high enough to be dry, it disappears. Now the Sarawak valley has this peculiarity—the lower portion though swampy, is not covered with a continuous lofty forest, but is principally occupied by the Nipa palm; and near the town of Sarawak where the country becomes dry, it is greatly undulated in many parts, and covered with small patches of virgin forest, and much second-growth jungle on the ground, which has once been cultivated by the Malays or Dyaks.

Now it seems probable to me that a wide extent of unbroken and equally lofty virgin forest is necessary to the comfortable existence of these animals. Such forests form their open country, where they can roam in every direction with as much facility as the Indian on the prairie, or the Arab on the desert, passing from tree-top to tree-top without ever being obliged to descend upon the earth. The elevated and the drier districts are more frequented by man, more cut up by clearings and low second-growth jungle—not adapted to its peculiar mode of progression, and where it would therefore be more exposed to danger, and more frequently obliged to descend upon the earth. There is probably also a greater variety of fruit in the Mias district, the small mountains which rise like islands out of it serving as gardens or plantations of a sort, where the trees of the uplands are to be found in the very midst of the swampy plains.

It is a singular and very interesting sight to watch a Mias making his way leisurely through the forest. He walks deliberately along some of the larger branches in the semi-erect attitude which the great length of his arms and the shortness of his legs cause him naturally to assume; and the disproportion between these limbs is increased by his walking on his knuckles, not on the palm of the hand, as we should do. He seems always to choose those branches which intermingle with an adjoining tree, on approaching which he stretches out his long arms, and seizing the opposing boughs, grasps them together with both hands, seems to try their strength, and then deliberately swings himself across to the next branch, on which he walks along as before. He never jumps or springs, or even appears to hurry himself, and yet manages to get along almost as quickly as a person can run through the forest beneath. The long and powerful arms are of the greatest use to the animal, enabling it to climb easily up the loftiest trees, to seize fruits and young leaves from slender boughs which will not bear its weight, and to gather leaves and branches with which to form its nest. I have already described how it forms a nest when wounded, but it uses a similar one to sleep on almost every night. This is placed low down, however, on a small tree not more than from twenty to fifty feet from the ground, probably because it is warmer and less exposed to wind than higher up. Each Mias is said to make a fresh one for himself every night; but I should think that is hardly probable, or their remains would be much more abundant; for though I saw several about the coal-mines, there must have been many Orangs about every day, and in a year their deserted nests would become very numerous. The Dyaks say that, when it is very wet, the Mias covers himself over with leaves of pandanus, or large ferns, which has perhaps led to the story of his making a hut in the trees.

The Orang does not leave his bed until the sun has well risen and has dried up the dew upon the leaves. He feeds all through the middle of the day, but seldom returns to the same tree two days running. They do not seem much alarmed at man, as they often stared down upon me for several minutes, and then only moved away slowly to an adjacent tree. After seeing one, I have often had to go half a mile or more to fetch my gun, and in nearly every case have found it on the same tree, or within a hundred yards, when I returned. I never saw two full-grown animals together, but both males and females are sometimes accompanied by half-grown young ones, while, at other times, three or four young ones were seen in company. Their food consists almost exclusively of fruit, with occasionally leaves, buds, and young shoots. They seem to prefer unripe fruits, some of which were very sour, others intensely bitter, particularly the large red, fleshy arillus of one which seemed an especial favourite. In other cases they eat only the small seed of a large fruit, and they almost always waste and destroy more than they eat, so that there is a continual rain of rejected portions below the tree they are feeding on. The Durian is an especial favourite, and quantities of this delicious fruit are destroyed wherever it grows surrounded by forest, but they will not cross clearings to get at them. It seems wonderful how the animal can tear open this fruit, the outer covering of which is so thick and tough, and closely covered with strong conical spines. It probably bites off a few of these first, and then, making a small hole, tears open the fruit with its powerful fingers.

The Mias rarely descends to the ground, except when pressed by hunger, it seeks succulent shoots by the riverside; or, in very dry weather, has to search after water, of which it generally finds sufficient in the hollows of leaves. Only once I saw two half-grown Orangs on the ground in a dry hollow at the foot of the Simunjon hill. They were playing together, standing erect, and grasping each other by the arms. It may be safely stated, however, that the Orang never walks erect, unless when using its hands to support itself by branches overhead or when attacked. Representations of its walking with a stick are entirely imaginary.

The Dyaks all declare that the Mias is never attacked by any animal in the forest, with two rare exceptions; and the accounts I received of these are so curious that I give them nearly in the words of my informants, old Dyak chiefs, who had lived all their lives in the places where the animal is most abundant. The first of whom I inquired said: "No animal is strong enough to hurt the Mias, and the only creature he ever fights with is the crocodile. When there is no fruit in the jungle, he goes to seek food on the banks of the river where there are plenty of young shoots that he likes, and fruits that grow close to the water. Then the crocodile sometimes tries to seize him, but the Mias gets upon him, and beats him with his hands and feet, and tears him and kills him."

He added that he had once seen such a fight, and that he believes that the Mias is always the victor.

My next informant was the Orang Kaya, or chief of the Balow Dyaks, on the Simunjon River. He said: "The Mias has no enemies; no animals dare attack it but the crocodile and the python. He always kills the crocodile by main strength, standing upon it, pulling open its jaws, and ripping up its throat. If a python attacks a Mias, he seizes it with his hands, and then bites it, and soon kills it. The Mias is very strong; there is no animal in the jungle so strong as he."

It is very remarkable that an animal so large, so peculiar, and of such a high type of form as the Orangutan, should be confined to so limited a district—to two islands, and those almost the last inhabited by the higher Mammalia; for, east of Borneo and Java, the Quadrumania, Ruminants, Carnivora, and many other groups of Mammalia diminish rapidly, and soon entirely disappear. When we consider, further, that almost all other animals have in earlier ages been represented by allied yet distinct forms—that, in the latter part of the tertiary period, Europe was inhabited by bears, deer, wolves, and cats; Australia by kangaroos and other marsupials; South America by gigantic sloths and anteaters; all different from any now existing, though intimately allied to them—we have every reason to believe that the Orangutan, the Chimpanzee, and the Gorilla have also had their forerunners. With what interest must every naturalist look forward to the time when the caves and tertiary deposits of the tropics may be thoroughly examined, and the past history and earliest appearance of the great man-like apes be made known at length.

I will now say a few words as to the supposed existence of a Bornean Orang as large as the Gorilla. I have myself examined the bodies of seventeen freshly-killed Orangs, all of which were carefully measured; and of seven of them, I preserved the skeleton. I also obtained two skeletons killed by other persons. Of this extensive series, sixteen were fully adult, nine being males, and seven females. The adult males of the large Orangs only varied from 4 feet 1 inch to 4 feet 2 inches in height, measured fairly to the heel, so as to give the height of the animal if it stood perfectly erect; the extent of the outstretched arms, from 7 feet 2 inches to 7 feet 8 inches; and the width of the face, from 10 inches to 13 1/2 inches. The dimensions given by other naturalists closely agree with mine. The largest Orang measured by Temminck was 4 feet high. Of twenty-five specimens collected by Schlegel and Muller, the largest old male was 4 feet 1 inch; and the largest skeleton in the Calcutta Museum was, according to Mr. Blyth, 4 feet 1 1/2 inch. My specimens were all from the northwest coast of Borneo; those of the Dutch from the west and south coasts; and no specimen has yet reached Europe exceeding these dimensions, although the total number of skins and skeletons must amount to over a hundred.

Strange to say, however, several persons declare that they have measured Orangs of a much larger size. Temminck, in his Monograph of the Orang, says that he has just received news of the capture of a specimen 5 feet 3 inches high. Unfortunately, it never seems to have a reached Holland, for nothing has since been heard of any such animal. Mr. St. John, in his "Life in the Forests of the Far East," vol. ii. p. 237, tells us of an Orang shot by a friend of his, which was 5 feet 2 inches from the heel to the top of the head, the arm 17 inches in girth, and the wrist 12 inches! The head alone was brought to Sarawak, and Mr. St. John tells us that he assisted to measure this, and that it was 15 inches broad by 14 long. Unfortunately, even this skull appears not to have been preserved, for no specimen corresponding to these dimensions has yet reached England.

In a letter from Sir James Brooke, dated October 1857 in which he acknowledges the receipt of my Papers on the Orang, published in the "Annals and Magazine of Natural History," he sends me the measurements of a specimen killed by his nephew, which I will give exactly as I received it: "September 3rd, 1867, killed female Orangutan. Height, from head to heel, 4 feet 6 inches. Stretch from fingers to fingers across body, 6 feet 1 inch. Breadth of face, including callosities, 11 inches." Now, in these dimensions, there is palpably one error; for in every Orang yet measured by any naturalist, an expanse of arms of 6 feet 1 inch corresponds to a height of about 3 feet 6 inches, while the largest specimens of 4 feet to 4 feet 2 inches high, always have the extended arms as much as 7 feet 3 inches to 7 feet 8 inches. It is, in fact, one of the characters of the genus to have the arms so long that an animal standing nearly erect can rest its fingers on the ground. A height of 4 feet 6 inches would therefore require a stretch of arms of at least 8 feet! If it were only 6 feet to that height, as given in the dimensions quoted, the animal would not be an Orang at all, but a new genus of apes, differing materially in habits and mode of progression. But Mr. Johnson, who shot this animal, and who knows Orangs well, evidently considered it to be one; and we have therefore to judge whether it is more probable that he made a mistake of two feet in the stretch of the arms, or of one foot in the height. The latter error is certainly the easiest to make, and it will bring his animal into agreement, as to proportions and size, with all those which exist in Europe. How easy it is to be deceived as to the height of these animals is well shown in the case of the Sumatran Orang, the skin of which was described by Dr. Clarke Abel. The captain and crew who killed this animal declared that when alive he exceeded the tallest man, and looked so gigantic that they thought he was 7 feet high; but that, when he was killed and lay upon the ground, they found he was only about 6 feet. Now it will hardly be credited that the skin of this identical animal exists in the Calcutta Museum, and Mr. Blyth, the late curator, states "that it is by no means one of the largest size"; which means that it is about 4 feet high!

Having these undoubted examples of error in the dimensions of Orangs, it is not too much to conclude that Mr. St. John's friend made a similar error of measurement, or rather, perhaps, of memory; for we are not told that the dimensions were noted down at the time they were made. The only figures given by Mr. St. John on his own authority are that "the head was 15 inches broad by 14 inches long." As my largest male was 13 1/2 broad across the face, measured as soon as the animal was killed, I can quite understand that when the head arrived at Sarawak from the Batang-Lupar, after two or three days' voyage, it was so swollen by decomposition as to measure an inch more than when it was fresh. On the whole, therefore, I think it will be allowed, that up to this time we have not the least reliable evidence of the existence of Orangs in Borneo more than 4 feet 2 inches high.

CHAPTER V. BORNEO—JOURNEY INTO THE INTERIOR.

(NOVEMBER 1855 TO JANUARY 1856.)

As the wet season was approaching, I determined to return to Sarawak, sending all my collections with Charles Allen around by sea, while I myself proposed to go up to the sources of the Sadong River and descend by the Sarawak valley. As the route was somewhat difficult, I took the smallest quantity of baggage, and only one servant, a Malay lad named Bujon, who knew the language of the Sadong Dyaks, with whom he had traded. We left the mines on the 27th of November, and the next day reached the Malay village of Gúdong, where I stayed a short time to buy fruit and eggs, and called upon the Datu Bandar, or Malay governor of the place. He lived in a large, and well-built house, very dirty outside and in, and was very inquisitive about my business, and particularly about the coal-mines. These puzzle the natives exceedingly, as they cannot understand the extensive and costly preparations for working coal, and cannot believe it is to be used only as fuel when wood is so abundant and so easily obtained. It was evident that Europeans seldom came here, for numbers of women skeltered away as I walked through the village and one girl about ten or twelve years old, who had just brought a bamboo full of water from the river, threw it down with a cry of horror and alarm the moment she caught sight of me, turned around and jumped into the stream. She swam beautifully, and kept looking back as if expecting I would follow her, screaming violently all the time; while a number of men and boys were laughing at her ignorant terror.

At Jahi, the next village, the stream became so swift in consequence of a flood, that my heavy boat could make no way, and I was obliged to send it back and go on in a very small open one. So far the river had been very monotonous, the banks being cultivated as rice-fields, and little thatched huts alone breaking the unpicturesque line of muddy bank crowned with tall grasses, and backed by the top of the forest behind the cultivated ground. A few hours beyond Jahi we passed the limits of cultivation, and had the beautiful virgin forest coming down to the water's edge, with its palms and creepers, its noble trees, its ferns, and epiphytes. The banks of the river were, however, still generally flooded, and we had some difficulty in finding a dry spot to sleep on. Early in the morning we reached Empugnan, a small Malay village, situated at the foot of an isolated mountain which had been visible from the mouth of the Simunjon River. Beyond here the tides are not felt, and we now entered upon a district of elevated forest, with a finer vegetation. Large trees stretch out their arms across the stream, and the steep, earthy banks are clothed with ferns and zingiberaceous plants.

Early in the afternoon we arrived at Tabókan, the first village of the Hill Dyaks. On an open space near the river, about twenty boys were playing at a game something like what we call "prisoner's base;" their ornaments of beads and brass wire and their gay-coloured kerchiefs and waist-cloths showing to much advantage, and forming a very pleasing sight. On being called by Bujon, they immediately left their game to carry my things up to the "headhouse,"—a circular building attached to most Dyak villages, and serving as a lodging for strangers, the place for trade, the sleeping-room of the unmarried youths, and the general council-chamber. It is elevated on lofty posts, has a large fireplace in the middle and windows in the roof all round, and forms a very pleasant and comfortable abode. In the evening it was crowded with young men and boys, who came to look at me. They were mostly fine young fellows, and I could not help admiring the simplicity and elegance of their costume. Their only dress is the long "chawat," or waist-cloth, which hangs down before and behind. It is generally of blue cotton, ending in three broad bands of red, blue, and white. Those who can afford it wear a handkerchief on the head, which is either red, with a narrow border of gold lace, or of three colours, like the "chawat." The large flat moon-shaped brass earrings, the heavy necklace of white or black beads, rows of brass rings on the arms and legs, and armlets of white shell, all serve to relieve and set off the pure reddish brown skin and jet-black hair. Add to this the little pouch containing materials for betel-chewing, and a long slender knife, both invariably worn at the side, and you have the everyday dress of the young Dyak gentleman.

The "Orang Kaya," or rich man, as the chief of the tribe is called, now came in with several of the older men; and the "bitchara" or talk commenced, about getting a boat and men to take me on the next morning. As I could not understand a word of their language, which is very different from Malay, I took no part in the proceedings, but was represented by my boy Bujon, who translated to me most of what was said. A Chinese trader was in the house, and he, too, wanted men the next day; but on his hinting this to the Orang Kaya, he was sternly told that a white man's business was now being discussed, and he must wait another day before his could be thought about.

After the "bitchara" was over and the old chiefs gone, I asked the young men to play or dance, or amuse themselves in their accustomed way; and after some little hesitation they agreed to do so. They first had a trial of strength, two boys sitting opposite each other, foot being placed against foot, and a stout stick grasped by both their hands. Each then tried to throw himself back, so as to raise his adversary up from the ground, either by main strength or by a sudden effort. Then one of the men would try his strength against two or three of the boys; and afterwards they each grasped their own ankle with a hand, and while one stood as firm as he could, the other swung himself around on one leg, so as to strike the other's free leg, and try to overthrow him. When these games had been played all around with varying success, we had a novel kind of concert. Some placed a leg across the knee, and struck the fingers sharply on the ankle, others beat their arms against their sides like a cock when he is going to crow, this making a great variety of clapping sounds, while another with his hand under his armpit produced a deep trumpet note; and, as they all kept time very well, the effect was by no means unpleasing. This seemed quite a favourite amusement with them, and they kept it up with much spirit.

The next morning we started in a boat about thirty feet long, and only twenty-eight inches wide. The stream here suddenly changes its character. Hitherto, though swift, it had been deep and smooth, and confined by steep banks. Now it rushed and rippled over a pebbly, sandy, or rocky bed, occasionally forming miniature cascades and rapids, and throwing up on one side or the other broad banks of finely coloured pebbles. No paddling could make way here, but the Dyaks with bamboo poles propelled us along with great dexterity and swiftness, never losing their balance in such a narrow and unsteady vessel, though standing up and exerting all their force. It was a brilliant day, and the cheerful exertions of the men, the rushing of the sparkling waters, with the bright and varied foliage, which from either bank stretched over our heads, produced an exhilarating sensation which recalled my canoe voyages on the grander waters of South America.

Early in the afternoon we reached the village of Borotói, and, though it would have been easy to reach the next one before night, I was obliged to stay, as my men wanted to return and others could not possibly go on with me without the preliminary talking. Besides, a white man was too great a rarity to be allowed to escape them, and their wives would never have forgiven them if, when they returned from the fields, they found that such a curiosity had not been kept for them to see. On entering the house to which I was invited, a crowd of sixty or seventy men, women, and children gathered around me, and I sat for half an hour like some strange animal submitted for the first time to the gaze of an inquiring public. Brass rings were here in the greatest profusion, many of the women having their arms completely covered with them, as well as their legs from the ankle to the knee. Round the waist they wear a dozen or more coils of fine rattan stained red, to which the petticoat is attached. Below this are generally a number of coils of brass wire, a girdle of small silver coins, and sometimes a broad belt of brass ring armour. On their heads they wear a conical hat without a crown, formed of variously coloured beads, kept in shape by rings of rattan, and forming a fantastic but not unpicturesque headdress.

Walking out to a small hill near the village, cultivated as a rice-field, I had a fine view of the country, which was becoming quite hilly, and towards the south, mountainous. I took bearings and sketches of all that was visible, an operation which caused much astonishment to the Dyaks who accompanied me, and produced a request to exhibit the compass when I returned. I was then surrounded by a larger crowd than before, and when I took my evening meal in the midst of a circle of about a hundred spectators anxiously observing every movement and criticising every mouthful, my thoughts involuntarily recurred to the lion at feeding time. Like those noble animals, I too was used to it, and it did not affect my appetite. The children here were more shy than at Tabókan, and I could not persuade them to play. I therefore turned showman myself, and exhibited the shadow of a dog's head eating, which pleased them so much that all the village in succession came out to see it. The "rabbit on the wall" does not do in Borneo, as there is no animal it resembles. The boys had tops shaped something like whipping-tops, but spun with a string.

The next morning we proceeded as before, but the river had become so rapid and shallow and the boats were all so small, that though I had nothing with me but a change of clothes, a gun, and a few cooking utensils, two were required to take me on. The rock which appeared here and there on the riverbank was an indurated clay-slate, sometimes crystalline, and thrown up almost vertically. Right and left of us rose isolated limestone mountains, their white precipices glistening in the sun and contrasting beautifully with the luxuriant vegetation that elsewhere clothed them. The river bed was a mass of pebbles, mostly pure white quartz, but with abundance of jasper and agate, presenting a beautifully variegated appearance. It was only ten in the morning when we arrived at Budw, and, though there were plenty of people about, I could not induce them to allow me to go on to the next village. The Orang Kaya said that if I insisted on having men, of course he would get them, but when I took him at his word and said I must have them, there came a fresh remonstrance; and the idea of my going on that day seemed so painful that I was obliged to submit. I therefore walked out over the rice-fields, which are here very extensive, covering a number of the little hills and valleys into which the whole country seems broken up, and obtained a fine view of hills and mountains in every direction.

In the evening the Orang Kaya came in full dress (a spangled velvet jacket, but no trousers), and invited me over to his house, where he gave me a seat of honour under a canopy of white calico and coloured handkerchiefs. The great verandah was crowded with people, and large plates of rice with cooked and fresh eggs were placed on the ground as presents for me. A very old man then dressed himself in bright-coloured cloths and many ornaments, and sitting at the door, murmured a long prayer or invocation, sprinkling rice from a basin he held in his hand, while several large gongs were loudly beaten and a salute of muskets fired off. A large jar of rice wine, very sour but with an agreeable flavour, was then handed around, and I asked to see some of their dances. These were, like most savage performances, very dull and ungraceful affairs; the men dressing themselves absurdly like women, and the girls making themselves as stiff and ridiculous as possible. All the time six or eight large Chinese gongs were being beaten by the vigorous arms of as many young men, producing such a deafening discord that I was glad to escape to the round house, where I slept very comfortably with half a dozen smoke-dried human skulls suspended over my head.

The river was now so shallow that boats could hardly get along. I therefore preferred walking to the next village, expecting to see something of the country, but was much disappointed, as the path lay almost entirely through dense bamboo thickets. The Dyaks get two crops off the ground in succession; one of rice, and the other of sugar-cane, maize, and vegetables. The ground then lies fallow eight or ten years, and becomes covered with bamboos and shrubs, which often completely arch over the path and shut out everything from the view. Three hours' walking brought us to the village of Senankan, where I was again obliged to remain the whole day, which I agreed to do on the promise of the Orang Kaya that his men should next day take me through two other villages across to Senna, at the head of the Sarawak River. I amused myself as I best could till evening, by walking about the high ground near, to get views of the country and bearings of the chief mountains. There was then another public audience, with gifts of rice and eggs, and drinking of rice wine. These Dyaks cultivate a great extent of ground, and supply a good deal of rice to Sarawak. They are rich in gongs, brass trays, wire, silver coins, and other articles in which a Dyak's wealth consists; and their women and children are all highly ornamented with bead necklaces, shells, and brass wire.

In the morning I waited some time, but the men that were to accompany me did not make their appearance. On sending to the Orang Kaya I found that both he and another head-man had gone out for the day, and on inquiring the reason was told that they could not persuade any of their men to go with me because the journey was a long and fatiguing one. As I was determined to get on, I told the few men that remained that the chiefs had behaved very badly, and that I should acquaint the Rajah with their conduct, and I wanted to start immediately. Every man present made some excuse, but others were sent for, and by dint of threats and promises, and the exertion of all Bujon's eloquence, we succeeded in getting off after two hours' delay.

For the first few miles our path lay over a country cleared for rice-fields, consisting entirely of small but deep and sharply-cut ridges and valleys without a yard of level ground. After crossing the Kayan river, a main branch of the Sadong, we got on to the lower slopes of the Seboran Mountain, and the path lay along a sharp and moderately steep ridge, affording an excellent view of the country. Its features were exactly those of the Himalayas in miniature, as they are described by Dr. Hooker and other travellers, and looked like a natural model of some parts of those vast mountains on a scale of about a tenth—thousands of feet being here represented by hundreds. I now discovered the source of the beautiful pebbles which had so pleased me in the riverbed. The slatey rocks had ceased, and these mountains seemed to consist of a sandstone conglomerate, which was in some places a mere mass of pebbles cemented together. I might have known that such small streams could not produce such vast quantities of well-rounded pebbles of the very hardest materials. They had evidently been formed in past ages, by the action of some continental stream or seabeach, before the great island of Borneo had risen from the ocean. The existence of such a system of hills and valleys reproducing in miniature all the features of a great mountain region, has an important bearing on the modern theory that the form of the ground is mainly due to atmospheric rather than to subterranean action. When we have a number of branching valleys and ravines running in many different directions within a square mile, it seems hardly possible to impute their formation, or even their origination, to rents and fissures produced by earthquakes. On the other hand, the nature of the rock, so easily decomposed and removed by water, and the known action of the abundant tropical rains, are in this case, at least, quite sufficient causes for the production of such valleys. But the resemblance between their forms and outlines, their mode of divergence, and the slopes and ridges that divide them, and those of the grand mountain scenery of the Himalayas, is so remarkable, that we are forcibly led to the conclusion that the forces at work in the two cases have been the same, differing only in the time they have been in action, and the nature of the material they have had to work upon.

About noon we reached the village of Menyerry, beautifully situated on a spur of the mountain about 600 feet above the valley, and affording a delightful view of the mountains of this part of Borneo. I here got a sight of Penrissen Mountain, at the head of the Sarawak River, and one of the highest in the district, rising to about 6,000 feet above the sea. To the south the Rowan, and further off the Untowan Mountains in the Dutch territory appeared equally lofty. Descending from Menyerry we again crossed the Kayan, which bends round the spur, and ascended to the pass which divides the Sadong and Sarawak valleys, and which is about 2,000 feet high. The descent from this point was very fine. A stream, deep in a rocky gorge, rushed on each side of us, to one of which we gradually descended, passing over many lateral gullys and along the faces of some precipices by means of native bamboo bridges. Some of these were several hundred feet long and fifty or sixty high, a single smooth bamboo four inches diameter forming the only pathway, while a slender handrail of the same material was often so shaky that it could only be used as a guide rather than a support.

Late in the afternoon we reached Sodos, situated on a spur between two streams, but so surrounded by fruit trees that little could be seen of the country. The house was spacious, clean and comfortable, and the people very obliging. Many of the women and children had never seen a white man before, and were very sceptical as to my being the same colour all over, as my face. They begged me to show them my arms and body, and they were so kind and good-tempered that I felt bound to give them some satisfaction, so I turned up my trousers and let them see the colour of my leg, which they examined with great interest.

In the morning early we continued our descent along a fine valley, with mountains rising 2,000 or 3,000 feet in every direction. The little river rapidly increased in size until we reached Senna, when it had become a fine pebbly stream navigable for small canoes. Here again the upheaved slatey rock appeared, with the same dip and direction as in the Sadong River. On inquiring for a boat to take me down the stream, I was told that the Senna Dyaks, although living on the river-banks, never made or used boats. They were mountaineers who had only come down into the valley about twenty years before, and had not yet got into new habits. They are of the same tribe as the people of Menyerry and Sodos. They make good paths and bridges, and cultivate much mountain land, and thus give a more pleasing and civilized aspect to the country than where the people move about only in boats, and confine their cultivation to the banks of the streams.

After some trouble I hired a boat from a Malay trader, and found three Dyaks who had been several times with Malays to Sarawak, and thought they could manage it very well. They turned out very awkward, constantly running aground, striking against rocks, and losing their balance so as almost to upset themselves and the boat—offering a striking contrast to the skill of the Sea Dyaks. At length we came to a really dangerous rapid where boats were often swamped, and my men were afraid to pass it. Some Malays with a boatload of rice here overtook us, and after safely passing down kindly sent back one of their men to assist me. As it was, my Dyaks lost their balance in the critical part of the passage, and had they been alone would certainly have upset the boat. The river now became exceedingly picturesque, the ground on each side being partially cleared for ricefields, affording a good view of the country. Numerous little granaries were built high up in trees overhanging the river, and having a bamboo bridge sloping up to them from the bank; and here and there bamboo suspension bridge crossed the stream, where overhanging trees favoured their construction.

I slept that night in the village of the Sebungow Dyaks, and the next day reached Sarawak, passing through a most beautiful country where limestone mountains with their fantastic forms and white precipices shot up on every side, draped and festooned with a luxuriant vegetation. The banks of the Sarawak River are everywhere covered with fruit trees, which supply the Dyaks with a great deal of their food. The Mangosteen, Lansat, Rambutan, Jack, Jambou, and Blimbing, are all abundant; but most abundant and most esteemed is the Durian, a fruit about which very little is known in England, but which both by natives and Europeans in the Malay Archipelago is reckoned superior to all others. The old traveller Linschott, writing in 1599, says: "It is of such an excellent taste that it surpasses in flavour all the other fruits of the world, according to those who have tasted it." And Doctor Paludanus adds: "This fruit is of a hot and humid nature. To those not used to it, it seems at first to smell like rotten onions, but immediately when they have tasted it, they prefer it to all other food. The natives give it honourable titles, exalt it, and make verses on it." When brought into a house the smell is often so offensive that some persons can never bear to taste it. This was my own case when I first tried it in Malacca, but in Borneo I found a ripe fruit on the ground, and, eating it out of doors, I at once became a confirmed Durian eater.

The Durian grows on a large and lofty forest tree, somewhat resembling an elm in its general character, but with a more smooth and scaly bark. The fruit is round or slightly oval, about the size of a large cocoanut, of a green colour, and covered all over with short stout spines the bases of which touch each other, and are consequently somewhat hexagonal, while the points are very strong and sharp. It is so completely armed, that if the stalk is broken off it is a difficult matter to lift one from the ground. The outer rind is so thick and tough, that from whatever height it may fall it is never broken. From the base to the apex five very faint lines may be traced, over which the spines arch a little; these are the sutures of the carpels, and show where the fruit may be divided with a heavy knife and a strong hand. The five cells are satiny white within, and are each filled with an oval mass of cream-coloured pulp, imbedded in which are two or three seeds about the size of chestnuts. This pulp is the eatable part, and its consistency and flavour are indescribable. A rich butter-like custard highly flavoured with almonds gives the best general idea of it, but intermingled with it come wafts of flavour that call to mind cream-cheese, onion-sauce, brown sherry, and other incongruities. Then there is a rich glutinous smoothness in the pulp which nothing else possesses, but which adds to its delicacy. It is neither acid, nor sweet, nor juicy; yet one feels the want of none of these qualities, for it is perfect as it is. It produces no nausea or other bad effect, and the more you eat of it the less you feel inclined to stop. In fact to eat Durians is a new sensation, worth a voyage to the East to experience.

When the fruit is ripe it falls of itself, and the only way to eat Durians in perfection is to get them as they fall; and the smell is then less overpowering. When unripe, it makes a very good vegetable if cooked, and it is also eaten by the Dyaks raw. In a good fruit season large quantities are preserved salted, in jars and bamboos, and kept the year round, when it acquires a most disgusting odour to Europeans, but the Dyaks appreciate it highly as a relish with their rice. There are in the forest two varieties of wild Durians with much smaller fruits, one of them orange-coloured inside; and these are probably the origin of the large and fine Durians, which are never found wild. It would not, perhaps, be correct to say that the Durian is the best of all fruits, because it cannot supply the place of the subacid juicy kinds, such as the orange, grape, mango, and mangosteen, whose refreshing and cooling qualities are so wholesome and grateful; but as producing a food of the most exquisite flavour, it is unsurpassed. If I had to fix on two only, as representing the perfection of the two classes, I should certainly choose the Durian and the Orange as the king and queen of fruits.

The Durian is, however, sometimes dangerous. When the fruit begins to ripen it falls daily and almost hourly, and accidents not unfrequently happen to persons walking or working under the trees. When a Durian strikes a man in its fall, it produces a dreadful wound, the strong spines tearing open the flesh, while the blow itself is very heavy; but from this very circumstance death rarely ensues, the copious effusion of blood preventing the inflammation which might otherwise take place. A Dyak chief informed me that he had been struck down by a Durian falling on his head, which he thought would certainly have caused his death, yet he recovered in a very short time.

Poets and moralists, judging from our English trees and fruits, have thought that small fruits always grew on lofty trees, so that their fall should be harmless to man, while the large ones trailed on the ground. Two of the largest and heaviest fruits known, however, the Brazil-nut fruit (Bertholletia) and Durian, grow on lofty forest trees, from which they fall as soon as they are ripe, and often wound or kill the native inhabitants. From this we may learn two things: first, not to draw general conclusions from a very partial view of nature; and secondly, that trees and fruits, no less than the varied productions of the animal kingdom, do not appear to be organized with exclusive reference to the use and convenience of man.

During my many journeys in Borneo, and especially during my various residences among the Dyaks, I first came to appreciate the admirable qualities of the Bamboo. In those parts of South America which I had previously visited, these gigantic grasses were comparatively scarce; and where found but little used, their place being taken as to one class of uses by the great variety of Palms, and as to another by calabashes and gourds. Almost all tropical countries produce Bamboos, and wherever they are found in abundance the natives apply them to a variety of uses. Their strength, lightness, smoothness, straightness, roundness and hollowness, the facility and regularity with which they can be split, their many different sizes, the varying length of their joints, the ease with which they can be cut and with which holes can be made through them, their hardness outside, their freedom from any pronounced taste or smell, their great abundance, and the rapidity of their growth and increase, are all qualities which render them useful for a hundred different purposes, to serve which other materials would require much more labour and preparation. The Bamboo is one of the most wonderful and most beautiful productions of the tropics, and one of nature's most valuable gifts to uncivilized man.

The Dyak houses are all raised on posts, and are often

two or three hundred feet long and forty or fifty wide. The floor is always formed of strips split from large Bamboos, so that each may be nearly flat and about three inches wide, and these are firmly tied down with rattan to the joists beneath. When well made, this is a delightful floor to walk upon barefooted, the rounded surfaces of the bamboo being very smooth and agreeable to the feet, while at the same time affording a firm hold. But, what is more important, they form with a mat over them an excellent bed, the elasticity of the Bamboo and its rounded surface being far superior to a more rigid and a flatter floor. Here we at once find a use for Bamboo which cannot be supplied so well by another material without a vast amount of labour—palms and other substitutes requiring much cutting and smoothing, and not being equally good when finished. When, however, a flat, close floor is required, excellent boards are made by splitting open large Bamboos on one side only, and flattening them out so as to form slabs eighteen inches wide and six feet long, with which some Dyaks floor their houses. These with constant rubbing of the feet and the smoke of years become dark and polished, like walnut or old oak, so that their real material can hardly be recognised. What labour is here saved to a savage whose only tools are an axe and a knife, and who, if he wants boards, must hew them out of the solid trunk of a tree, and must give days and weeks of labour to obtain a surface as smooth and beautiful as the Bamboo thus treated affords him. Again, if a temporary house is wanted, either by the native in his plantation or by the traveller in the forest, nothing is so convenient as the Bamboo, with which a house can be constructed with a quarter of the labour and time than if other materials are used.

As I have already mentioned, the Hill Dyaks in the interior of Sarawak make paths for long distances from village to village and to their cultivated grounds, in the course of which they have to cross many gullies and ravines, and even rivers; or sometimes, to avoid a long circuit, to carry the path along the face of a precipice. In all these cases the bridges they construct are of Bamboo, and so admirably adapted is the material for this purpose, that it seems doubtful whether they ever would have attempted such works if they had not possessed it. The Dyak bridge is simple but well designed. It consists merely of stout Bamboos crossing each other at the road-way like the letter X, and rising a few feet above it. At the crossing they are firmly bound together, and to a large Bamboo which lays upon them and forms the only pathway, with a slender and often very shaky one to serve as a handrail. When a river is to be crossed, an overhanging tree is chosen from which the bridge is partly suspended and partly supported by diagonal struts from the banks, so as to avoid placing posts in the stream itself, which would be liable to be carried away by floods. In carrying a path along the face of a precipice, trees and roots are made use of for suspension; struts arise from suitable notches or crevices in the rocks, and if these are not sufficient, immense Bamboos fifty or sixty feet long are fixed on the banks or on the branch of a tree below. These bridges are traversed daily by men and women carrying heavy loads, so that any insecurity is soon discovered, and, as the materials are close at hand, immediately repaired. When a path goes over very steep ground, and becomes slippery in very wet or very dry weather, the Bamboo is used in another way. Pieces are cut about a yard long, and opposite notches being made at each end, holes are formed through which pegs are driven, and firm and convenient steps are thus formed with the greatest ease and celerity. It is true that much of this will decay in one or two seasons, but it can be so quickly replaced as to make it more economical than using a harder and more durable wood.

One of the most striking uses to which Bamboo is applied by the Dyaks, is to assist them in climbing lofty trees by driving in pegs in the way I have already described at page 85. This method is constantly used in order to obtain wax, which is one of the most valuable products of the country. The honey-bee of Borneo very generally hangs its combs under the branches of the Tappan, a tree which towers above all others in the forest, and whose smooth cylindrical trunk often rises a hundred feet without a branch. The Dyaks climb these lofty trees at night, building up their Bamboo ladder as they go, and bringing down gigantic honeycombs. These furnish them with a delicious feast of honey and young bees, besides the wax, which they sell to traders, and with the proceeds buy the much-coveted brass wire, earrings, and bold-edged handkerchiefs with which they love to decorate themselves. In ascending Durian and other fruit trees which branch at from thirty to fifty feet from the ground, I have seen them use the Bamboo pegs only, without the upright Bamboo which renders them so much more secure.

The outer rind of the Bamboo, split and shaved thin, is the strongest material for baskets; hen-coops, bird-cages, and conical fish-traps are very quickly made from a single joint, by splitting off the skin in narrow strips left attached to one end, while rings of the same material or of rattan are twisted in at regular distances. Water is brought to the houses by little aqueducts formed of large Bamboos split in half and supported on crossed sticks of various heights so as to give it a regular fall. Thin long-jointed Bamboos form the Dyaks' only water-vessels, and a dozen of them stand in the corner of every house. They are clean, light, and easily carried, and are in many ways superior to earthen vessels for the same purpose. They also make excellent cooking utensils; vegetables and rice can be boiled in them to perfection, and they are often used when travelling. Salted fruit or fish, sugar, vinegar, and honey are preserved in them instead of in jars or bottles. In a small Bamboo case, prettily carved and ornamented, the Dyak carries his sirih and lime for betel chewing, and his little long-bladed knife has a Bamboo sheath. His favourite pipe is a huge hubble-bubble, which he will construct in a few minutes by inserting a small piece of Bamboo for a bowl obliquely into a large cylinder about six inches from the bottom containing water, through which the smoke passes to a long slender Bamboo tube. There are many other small matters for which Bamboo is daily used, but enough has now been mentioned to show its value. In other parts of the Archipelago I have myself seen it applied to many new uses, and it is probable that my limited means of observation did not make me acquainted with one-half the ways in which it is serviceable to the Dyaks of Sarawak.

While upon the subject of plants I may here mention a few of the more striking vegetable productions of Borneo. The wonderful Pitcher-plants, forming the genus Nepenthes of botanists, here reach their greatest development. Every mountain-top abounds with them, running along the ground, or climbing over shrubs and stunted trees; their elegant pitchers hanging in every direction. Some of these are long and slender, resembling in form the beautiful Philippine lace-sponge (Euplectella), which has now become so common; others are broad and short. Their colours are green, variously tinted and mottled with red or purple. The finest yet known were obtained on the summit of Kini-balou, in North-west Borneo. One of the broad sort, Nepenthes rajah, will hold two quarts of water in its pitcher. Another, Nepenthes Edwardsiania, has a narrow pitcher twenty inches long; while the plant itself grows to a length of twenty feet.

Ferns are abundant, but are not so varied as on the volcanic mountains of Java; and Tree-ferns are neither so plentiful nor so large as on that island. They grow, however, quite down to the level of the sea, and are generally slender and graceful plants from eight to fifteen feet high. Without devoting much time to the search I collected fifty species of Ferns in Borneo, and I have no doubt a good botanist would have obtained twice the number. The interesting group of Orchids is very abundant, but, as is generally the case, nine-tenths of the species have small and inconspicuous flowers. Among the exceptions are the fine Coelogynes, whose large clusters of yellow flowers ornament the gloomiest forests, and that most extraordinary plant, Vanda Lowii, which last is particularly abundant near some hot springs at the foot of the Penin-jauh Mountain. It grows on the lower branches of trees, and its strange pendant flower-spires often hang down so as almost to reach the ground. These are generally six or eight feet long, bearing large and handsome flowers three inches across, and varying in colour from orange to red, with deep purple-red spots. I measured one spike, which reached the extraordinary length of nine feet eight inches, and bore thirty-six flowers, spirally arranged upon a slender thread-like stalk. Specimens grown in our English hot-houses have produced flower-spires of equal length, and with a much larger number of blossoms.

Flowers were scarce, as is usual in equatorial forests, and it was only at rare intervals that I met with anything striking. A few fine climbers were sometimes seen, especially a handsome crimson and yellow Aeschynanthus, and a fine leguminous plant with clusters of large Cassia-like flowers of a rich purple colour. Once I found a number of small Anonaceous trees of the genus Polyalthea, producing a most striking effect in the gloomy forest shades. They were about thirty feet high, and their slender trunks were covered with large star-like crimson flowers, which clustered over them like garlands, and resembled some artificial decoration more than a natural product.

The forests abound with gigantic trees with cylindrical, buttressed, or furrowed stems, while occasionally the traveller comes upon a wonderful fig-tree, whose trunk is itself a forest of stems and aerial roots. Still more rarely are found trees which appear to have begun growing in mid-air, and from the same point send out wide-spreading branches above and a complicated pyramid of roots descending for seventy or eighty feet to the ground below, and so spreading on every side, that one can stand in the very centre with the trunk of the tree immediately overhead. Trees of this character are found all over the Archipelago, and the accompanying illustration (taken from one which I often visited in the Aru

Islands) will convey some idea of their general character. I believe that they originate as parasites, from seeds carried by birds and dropped in the fork of some lofty tree. Hence descend aerial roots, clasping and ultimately destroying the supporting tree, which is in time entirely replaced by the humble plant which was at first dependent upon it. Thus we have an actual struggle for life in the vegetable kingdom, not less fatal to the vanquished than the struggles among animals which we can so much more easily observe and understand. The advantage of quicker access to light and warmth and air, which is gained in one way by climbing plants, is here obtained by a forest tree, which has the means of starting in life at an elevation which others can only attain after many years of growth, and then only when the fall of some other tree has made room for then. Thus it is that in the warm and moist and equable climate of the tropics, each available station is seized upon and becomes the means of developing new forms of life especially adapted to occupy it.

On reaching Sarawak early in December, I found there would not be an opportunity of returning to Singapore until the latter end of January. I therefore accepted Sir James Brooke's invitation to spend a week with him and Mr. St. John at his cottage on Peninjauh. This is a very steep pyramidal mountain of crystalline basaltic rock, about a thousand feet high, and covered with luxuriant forest. There are three Dyak villages upon it, and on a little platform near the summit is the rude wooden lodge where the English Rajah was accustomed to go for relaxation and cool fresh air. It is only twenty miles up the river, but the road up the mountain is a succession of ladders on the face of precipices, bamboo bridges over gullies and chasms, and slippery paths over rocks and tree-trunks and huge boulders as big as houses. A cool spring under an overhanging rock just below the cottage furnished us with refreshing baths and delicious drinking water, and the Dyaks brought us daily heaped-up baskets of Mangosteens and Lansats, two of the most delicious of the subacid tropical fruits. We returned to Sarawak for Christmas (the second I had spent with Sir James Brooke), when all the Europeans both in the town and from the out-stations enjoyed the hospitality of the Rajah, who possessed in a pre-eminent degree the art of making every one around him comfortable and happy.

A few days afterwards I returned to the mountain with Charles and a Malay boy named Ali and stayed there three weeks for the purpose of making a collection of land-shells, butterflies and moths, ferns and orchids. On the hill itself ferns were tolerably plentiful, and I made a collection of about forty species. But what occupied me most was the great abundance of moths which on certain occasions I was able to capture. As during the whole of my eight years' wanderings in the East I never found another spot where these insects were at all plentiful, it will be interesting to state the exact conditions under which I here obtained them.

On one side of the cottage there was a verandah, looking down the whole side of the mountain and to its summit on the right, all densely clothed with forest. The boarded sides of the cottage were whitewashed, and the roof of the verandah was low, and also boarded and whitewashed. As soon as it got dark I placed my lamp on a table against the wall, and with pins, insect-forceps, net, and collecting-boxes by my side, sat down with a book. Sometimes during the whole evening only one solitary moth would visit me, while on other nights they would pour in, in a continual stream, keeping me hard at work catching and pinning till past midnight. They came literally by the thousands. These good nights were very few. During the four weeks that I spent altogether on the hill I only had four really good nights, and these were always rainy, and the best of them soaking wet. But wet nights were not always good, for a rainy moonlight night produced next to nothing. All the chief tribes of moths were represented, and the beauty and variety of the species was very great. On good nights I was able to capture from a hundred to two hundred and fifty moths, and these comprised on each occasion from half to two-thirds that number of distinct species. Some of them would settle on the wall, some on the table, while many would fly up to the roof and give me a chase all over the verandah before I could secure them. In order to show the curious connection between the state of weather and the degree in which moths were attracted to light, I add a list of my captures each night of my stay on the hill:

Date (1855)	No. of Moths	Remarks
Dec. 13th	1	Fine; starlight.
14th	75	Drizzly and fog.
15th	41	Showery; cloudy.
16th	158	(120 species.) Steady rain.
17th	82	Wet; rather moonlight.
18th	9	Fine; moonlight.
19th	2	Fine; clear moonlight.
31st	200	(130 species.) Dark and windy; heavy rain.

Date (1856)		
Jan. 1st	185	Very wet.
2d	68	Cloudy and showers.
3d	50	Cloudy.
4th	12	Fine.
5th	10	Fine.
6th	8	Very fine.
7th	8	Very fine.
8th	10	Fine.
9th	36	Showery.
10th	30	Showery.

11th	260	Heavy rain all night, and dark.
12th	56	Showery.
13th	44	Showery; some moonlight.
14th	4	Fine; moonlight.
15th	24	Rain; moonlight.
16th	6	Showers; moonlight.
17th	6	Showers; moonlight.
18th	1	Showers; moonlight.
Total	1,386	

It thus appears that on twenty-six nights I collected 1,386 moths, but that more than 800 of them were collected on four very wet and dark nights. My success here led me to hope that, by similar arrangements, I might on every island be able to obtain an abundance of these insects; but, strange to say, during the six succeeding years, I was never once able to make any collections at all approaching those at Sarawak. The reason for this I can pretty well understand to be owing to the absence of some one or other essential condition that were here all combined. Sometimes the dry season was the hindrance; more frequently residence in a town or village not close to virgin forest, and surrounded by other houses whose lights were a counter-attraction; still more frequently residence in a dark palm-thatched house, with a lofty roof, in whose recesses every moth was lost the instant it entered. This last was the greatest drawback, and the real reason why

I never again was able to make a collection of moths; for I never afterwards lived in a solitary jungle-house with a low boarded and whitewashed verandah, so constructed as to prevent insects at once escaping into the upper part of the house, quite out of reach.

After my long experience, my numerous failures, and my one success, I feel sure that if any party of naturalists ever make a yacht-voyage to explore the Malayan Archipelago, or any other tropical region, making entomology one of their chief pursuits, it would well repay them to carry a small framed verandah, or a verandah-shaped tent of white canvas, to set up in every favourable situation, as a means of making a collection of nocturnal Lepidoptera, and also of obtaining rare specimens of Coleoptera and other insects. I make the suggestion here, because no one would suspect the enormous difference in results that such an apparatus would produce; and because I consider it one of the curiosities of a collector's experience, to have found out that some such apparatus is required.

When I returned to Singapore I took with me the Malay lad named Ali, who subsequently accompanied me all over the Archipelago. Charles Allen preferred staying at the Mission-house, and afterwards obtained employment in Sarawak and in Singapore, until he again joined me four years later at Amboyna in the Moluccas.

CHAPTER VI. BORNEO—THE DYAKS.

THE manners and customs of the aborigines of Borneo have been described in great detail, and with much fuller information than I possess, in the writings of Sir James Brooke, Messrs. Low, St. John, Johnson Brooke, and many others. I do not propose to go over the ground again, but shall confine myself to a sketch, from personal observation, of the general character of the Dyaks, and of such physical, moral, and social characteristics as have been less frequently noticed.

The Dyak is closely allied to the Malay, and more remotely to the Siamese, Chinese, and other Mongol races. All these are characterised by a reddish-brown or yellowish-brown skin of various shades, by jet-black straight hair, by the scanty or deficient beard, by the rather small and broad nose, and high cheekbones; but none of the Malayan races have the oblique eyes which are characteristic of the more typical Mongols. The average stature of the Dyaks is rather more than that of the Malays, while it is considerably under that of most Europeans. Their forms are well proportioned, their feet and hands small, and they rarely or never attain the bulk of body so often seen in Malays and Chinese.

I am inclined to rank the Dyaks above the Malays

in mental capacity, while in moral character they are undoubtedly superior to them. They are simple and honest, and become the prey of the Malay and Chinese traiders, who cheat and plunder them continually. They are more lively, more talkative, less secretive, and less suspicious than the Malay, and are therefore pleasanter companions. The Malay boys have little inclination for active sports and games, which form quite a feature in the life of the Dyak youths, who, besides outdoor games of skill and strength, possess a variety of indoor amusements. One wet day, in a Dyak house, when a number of boys and young men were about me, I thought to amuse them with something new, and showed them how to make "cat's cradle" with a piece of string. Greatly to my surprise, they knew all about it, and more than I did; for, after Charles and I had gone through all the changes we could make, one of the boys took it off my hand, and made several new figures which quite puzzled me. They then showed me a number of other tricks with pieces of string, which seemed a favourite amusement with them.

Even these apparently trifling matters may assist us to form a truer estimate of the Dyaks' character and social condition. We learn thereby, that these people have passed beyond that first stage of savage life in which the struggle for existence absorbs all of the faculties, and in which every thought and idea is connected with war or hunting, or the provision for their immediate necessities. These amusements indicate a capability of civilization, an aptitude to enjoy other than mere sensual pleasures, which might be taken advantage of to elevate their whole intellectual and social life.

The moral character of the Dyaks is undoubtedly high—a statement which will seem strange to those who have heard of them only as head-hunters and pirates. The Hill Dyaks of whom I am speaking, however, have never been pirates, since they never go near the sea; and head-hunting is a custom originating in the petty wars of village with village, and tribe with tribe, which no more implies a bad moral character than did the custom of the slave-trade a hundred years ago imply want of general morality in all who participated in it. Against this one stain on their character (which in the case of the Sarawak Dyaks no longer exists) we have to set many good points. They are truthful and honest to a remarkable degree. From this cause it is very often impossible to get from them any definite information, or even an opinion. They say, "If I were to tell you what I don't know, I might tell a lie;" and whenever they voluntarily relate any matter of fact, you may be sure they are speaking the truth. In a Dyak village the fruit trees have each their owner, and it has often happened to me, on asking an inhabitant to gather me some fruit, to be answered, "I can't do that, for the owner of the tree is not here;" never seeming to contemplate the possibility of acting otherwise. Neither will they take the smallest thing belonging to an European. When living at Simunjon, they continually came to my house, and would pick up scraps of torn newspaper or crooked pins that I had thrown away, and ask as a great favour whether they might have them. Crimes of violence (other than head-hunting) are almost unknown; for in twelve years, under Sir James Brooke's rule, there had been only one case of murder in a Dyak tribe, and that one was committed by a stranger who had been adopted into the tribe. In several other matters of morality they rank above most uncivilized, and even above many civilized nations. They are temperate in food and drink, and the gross sensuality of the Chinese and Malays is unknown among them. They have the usual fault of all people in a half-savage state—apathy and dilatoriness, but, however annoying this may be to Europeans who come in contact with them, it cannot be considered a very grave offence, or be held to outweigh their many excellent qualities.

During my residence among the Hill Dyaks, I was much struck by the apparent absence of those causes which are generally supposed to check the increase of population, although there were plain indications of stationary or but slowly increasing numbers. The conditions most favourable to a rapid increase of population are: an abundance of food, a healthy climate, and early marriages. Here these conditions all exist. The people produce far more food than they consume, and exchange the surplus for gongs and brass cannon, ancient jars, and gold and silver ornaments, which constitute their wealth. On the whole, they appear very free from disease, marriages take place early (but not too early), and old bachelors and old maids are alike unknown. Why, then, we must inquire, has not a greater population been produced? Why are the Dyak villages so small and so widely scattered, while nine-tenths of the country is still covered with forest?

Of all the checks to population among savage nations mentioned by Malthus—starvation, disease, war, infanticide, immorality, and infertility of the women—the last is that which he seems to think least important, and of doubtful efficacy; and yet it is the only one that seems to me capable of accounting for the state of the population among the Sarawak Dyaks. The population of Great Britain increases so as to double itself in about fifty years. To do this it is evident that each married couple must average three children who live to be married at the age of about twenty-five. Add to these those who die in infancy, those who never marry, or those who marry late in life and have no offspring, the number of children born to each marriage must average four or five, and we know that families of seven or eight are very common, and of ten and twelve by no means rare. But from inquiries at almost every Dyak tribe I visited, I ascertained that the women rarely had more than three or four children, and an old chief assured me that he had never known a woman to have more than seven.

In a village consisting of a hundred and fifty families, only one consisted of six children living, and only six of five children, the majority of families appearing to be two, three, or four. Comparing this with the known proportions in European countries, it is evident that the number of children to each marriage can hardly average more than three or four; and as even in civilized countries half the population die before the age of twenty-five, we should have only two left to replace their parents; and so long as this state of things continued, the population must remain stationary. Of course this is a mere illustration; but the facts I have stated seem to indicate that something of the kind really takes place; and if so, there is no difficulty in understanding the smallness and almost stationary population of the Dyak tribes.

We have next to inquire what is the cause of the small number of births and of living children in a family. Climate and race may have something to do with this, but a more real and efficient cause seems to me to be the hard labour of the women, and the heavy weights they constantly carry. A Dyak woman generally spends the whole day in the field, and carries home every night a heavy load of vegetables and firewood, often for several miles, over rough and hilly paths; and not unfrequently has to climb up a rocky mountain by ladders, and over slippery stepping-stones, to an elevation of a thousand feet. Besides this, she has an hour's work every evening to pound the rice with a heavy wooden stamper, which violently strains every part of the body. She begins this kind of labour when nine or ten years old, and it never ceases but with the extreme decrepitude of age. Surely we need not wonder at the limited number of her progeny, but rather be surprised at the successful efforts of nature to prevent the extermination of the race.

One of the surest and most beneficial effects of advancing civilization, will be the amelioration of the condition of these women. The precept and example of higher races will make the Dyak ashamed of his comparatively idle life, while his weaker partner labours like a beast of burthen. As his wants become increased and his tastes refined, the women will have more household duties to attend to, and will then cease to labour in the field—a change which has already to a great extent taken place in the allied Malay, Javanese, and Bugis tribes. Population will then certainly increase more rapidly, improved systems of agriculture and some division of labour will become necessary in order to provide the means of existence, and a more complicated social state will take the place of the simple conditions of society which now occur among them. But, with the sharper struggle for existence that will then arise, will the happiness of the people as a whole be increased or diminished? Will not evil passions be aroused by the spirit of competition, and crimes and vices, now unknown or dormant, be called into active existence? These are problems that time alone can solve; but it is to be hoped that education and a high-class European example may obviate much of the evil that too often arises in analogous cases, and that we may at length be able to point to one instance of an uncivilized people who have not become demoralized, and finally exterminated, by contact with European civilization.

A few words in conclusion, about the government of Sarawak. Sir James Brooke found the Dyaks oppressed and ground down by the most cruel tyranny. They were cheated by the Malay traders and robbed by the Malay chiefs. Their wives and children were often captured and sold into slavery, and hostile tribes purchased permission from their cruel rulers to plunder, enslave, and murder them. Anything like justice or redress for these injuries was utterly unattainable. From the time Sir James obtained possession of the country, all this was stopped. Equal justice was awarded to Malay, Chinaman, and Dyak. The remorseless pirates from the rivers farther east were punished, and finally shut up within their own territories, and the Dyak, for the first time, could sleep in peace. His wife and children were now safe from slavery; his house was no longer burned over his head; his crops and his fruits were now his own to sell or consume as he pleased. And the unknown stranger who had done all this for them, and asked for nothing in return, what could he be? How was it possible for them to realize his motives? Was it not natural that they should refuse to believe he was a man? For of pure benevolence combined with great power, they had had no experience among men. They naturally concluded that he was a superior being, come down upon earth to confer blessings on the afflicted. In many villages where he had not been seen, I was asked strange questions about him. Was he not as old as the mountains? Could he not bring the dead to life? And they firmly believe that he can give them good harvests, and make their fruit-trees bear an abundant crop.

In forming a proper estimate of Sir James Brooke's government it must ever be remembered that he held Sarawak solely by the goodwill of the native inhabitant. He had to deal with two races, one of whom, the Mahometan Malays, looked upon the other race, the Dyaks, as savages and slaves, only fit to be robbed and plundered. He has effectually protected the Dyaks, and has invariably treated them as, in his sight, equal to the Malays; and yet he has secured the affection and goodwill of both. Notwithstanding the religious prejudice, of Mahometans, he has induced them to modify many of their worst laws and customs, and to assimilate their criminal code to that of the civilized world. That his government still continues, after twenty-seven years—notwithstanding his frequent absences from ill-health, notwithstanding conspiracies of Malay chiefs, and insurrections of Chinese gold-diggers, all of which have been overcome by the support of the native population, and notwithstanding financial, political, and domestic troubles is due, I believe, solely to the many admirable qualities which Sir James Brooke possessed, and especially to his having convinced the native population, by every action of his life, that he ruled them, not for his own advantage, but for their good.

Since these lines were written, his noble spirit has passed away. But though, by those who knew him not, he may be sneered at as an enthusiastic adventurer, abused as a hardhearted despot, the universal testimony of everyone who came in contact with him in his adopted country, whether European, Malay, or Dyak, will be, that Rajah Brooke was a great, a wise, and a good ruler; a true and faithful friend—a man to be admired for his talents, respected for his honesty and courage, and loved for his genuine hospitality, his kindness of disposition, and his tenderness of heart.

CHAPTER VII. JAVA.

I SPENT three months and a half in Java, from July 18th to October 31st, 1861, and shall briefly describe my own movements, and my observations of the people and the natural history of the country. To all those who wish to understand how the Dutch now govern Java, and how it is that they are enabled to derive a large annual revenue from it, while the population increases, and the inhabitants are contented, I recommend the study of Mr. Money's excellent and interesting work, "How to Manage a Colony." The main facts and conclusions of that work I most heartily concur in, and I believe that the Dutch system is the very best that can be adopted, when a European nation conquers or otherwise acquires possession of a country inhabited by an industrious but semi-barbarous people. In my account of Northern Celebes, I shall show how successfully the same system has been applied to a people in a very different state of civilization from the Javanese; and in the meanwhile will state in the fewest words possible what that system is.

The mode of government now adopted in Java is to retain the whole series of native rulers, from the village chief up to princes, who, under the name of Regents, are the heads of districts about the size of a small English county.

With each Regent is placed a Dutch Resident, or Assistant Resident, who is considered to be his "elder brother," and whose "orders" take the form of "recommendations," which are, however, implicitly obeyed. Along with each Assistant Resident is a Controller, a kind of inspector of all the lower native rulers, who periodically visits every village in the district, examines the proceedings of the native courts, hears complaints against the head-men or other native chiefs, and superintends the Government plantations. This brings us to the "culture system," which is the source of all the wealth the Dutch derive from Java, and is the subject of much abuse in this country because it is the reverse of "free trade." To understand its uses and beneficial effects, it is necessary first to sketch the common results of free European trade with uncivilized peoples.

Natives of tropical climates have few wants, and, when these are supplied, are disinclined to work for superfluities without some strong incitement. With such a people the introduction of any new or systematic cultivation is almost impossible, except by the despotic orders of chiefs whom they have been accustomed to obey, as children obey their parents. The free competition of European traders, however introduces two powerful inducements to exertion. Spirits or opium is a temptation too strong for most savages to resist, and to obtain these he will sell whatever he has, and will work to get more. Another temptation he cannot resist, is goods on credit. The trader offers him gay cloths, knives, gongs, guns, and gunpowder, to be paid for by some crop perhaps not yet planted, or some product yet in the forest. He has not sufficient forethought to take only a moderate quantity, and not enough energy to work early and late in order to get out of debt; and the consequence is that he accumulates debt upon debt, and often remains for years, or for life, a debtor and almost a slave. This is a state of things which occurs very largely in every part of the world in which men of a superior race freely trade with men of a lower race. It extends trade no doubt for a time, but it demoralizes the native, checks true civilization—and does not lead to any permanent increase in the wealth of the country; so that the European government of such a country must be carried on at a loss.

The system introduced by the Dutch was to induce the people, through their chiefs, to give a portion of their time, to the cultivation of coffee, sugar, and other valuable products. A fixed rate of wages—low indeed, but, about equal to that of all places where European competition has not artificially raised it—was paid to the labourers engaged in clearing the ground and forming the plantations under Government superintendence. The produce is sold to the Government at a low, fixed price. Out of the net profit a percentage goes to the chiefs, and the remainder is divided among the workmen. This surplus in good years is something considerable. On the whole, the people are well fed and decently clothed, and have acquired habits of steady industry and the art of scientific cultivation, which must be of service to them in the future. It must be remembered, that the Government expended capital for years before any return was obtained; and if they now derive a large revenue, it is in a way which is far less burthensome, and far more beneficial to the people, than any tax that could be levied.

But although the system may be a good one, and as well adapted to the development of arts and industry in a half civilized people as it is to the material advantage of the governing country, it is not pretended that in practice it is perfectly carried out. The oppressive and servile relations between chiefs and people, which have continued for perhaps a thousand years, cannot be at once abolished; and some evil must result from those relations, until the spread of education and the gradual infusion of European blood causes it naturally and insensibly to disappear. It is said that the Residents, desirous of showing a large increase in the products of their districts, have sometimes pressed the people to such continued labour on the plantations that their rice crops have been materially diminished, and famine has been the result. If this has happened, it is certainly not a common thing, and is to be set down to the abuse of the system, by the want of judgment, or want of humanity in the Resident.

Volume I

A tale has lately been written in Holland, and translated into English, entitled "Max Havelaar;" or, the "Coffee Auctions of the Dutch Trading Company," and with our usual one-sidedness in all relating to the Dutch Colonial System, this work has been excessively praised, both for its own merits, and for its supposed crushing exposure of the iniquities of the Dutch government of Java. Greatly to my surprise, I found it a very tedious and long-winded story, full of rambling digressions; and whose only point is to show that the Dutch Residents and Assistant Residents wink at the extortions of the native princes; and that in some districts the natives have to do work without payment, and have their goods taken away from them without compensation. Every statement of this kind is thickly interspersed with italics and capital letters; but as the names are all fictitious, and neither dates, figures, nor details are ever given, it is impossible to verify or answer them. Even if not exaggerated, the facts stated are not nearly so bad as those of the oppression by free-trade indigo-planters, and torturing by native tax-gatherers under British rule in India, with which the readers of English newspapers were familiar a few years ago. Such oppression, however, is not fairly to be imputed in either case to the particular form of government, but is rather due to the infirmity of human nature, and to the impossibility of at once destroying all trace of ages of despotism on the one side, and of slavish obedience to their chiefs on the other.

It must be remembered, that the complete establishment of the Dutch power in Java is much more recent than that of our rule in India, and that there have been several changes of government, and in the mode of raising revenue. The inhabitants have been so recently under the rule of their native princes, that it is not easy at once to destroy the excessive reverence they feel for their old masters, or to diminish the oppressive exactions which the latter have always been accustomed to make. There is, however, one grand test of the prosperity, and even of the happiness, of a community, which we can apply here—the rate of increase of the population.

It is universally admitted that when a country increases rapidly in population, the people cannot be very greatly oppressed or very badly governed. The present system of raising a revenue by the cultivation of coffee and sugar, sold to Government at a fixed price, began in 1832. Just before this, in 1826, the population by census was 5,500,000, while at the beginning of the century it was estimated at 3,500,000. In 1850, when the cultivation system had been in operation eighteen years, the population by census was over 9,500,000, or an increase of 73 per cent in twenty-four years. At the last census, in 1865, it amounted to 14,168,416, an increase of very nearly 50 per cent in fifteen years—a rate which would double the population in about twenty-six years. As Java (with Madura) contains about 38,500 geographical square miles, this will give an average of 368 persons to the square mile, just double that of the populous and fertile Bengal Presidency as given in Thornton's Gazetteer of India, and fully one-third more than that of Great Britain and Ireland at the last Census. If, as I believe, this vast population is on the whole contented and happy, the Dutch Government should consider well before abruptly changing a system which has led to such great results.

Taking it as a whole, and surveying it from every point of view, Java is probably the very finest and most interesting tropical island in the world. It is not first in size, but it is more than 600 miles long, and from 60 to 120 miles wide, and in area is nearly equal to England; and it is undoubtedly the most fertile, the most productive, and the most populous island within the tropics. Its whole surface is magnificently varied with mountain and forest scenery. It possesses thirty-eight volcanic mountains, several of which rise to ten or twelve thousand feet high. Some of these are in constant activity, and one or other of them displays almost every phenomenon produced by the action of subterranean fires, except regular lava streams, which never occur in Java. The abundant moisture and tropical heat of the climate causes these mountains to be clothed with luxuriant vegetation, often to their very summits, while forests and plantations cover their lower slopes. The animal productions, especially the birds and insects, are beautiful and varied, and present many peculiar forms found nowhere else upon the globe.

The soil throughout the island is exceedingly fertile, and all the productions of the tropics, together with many of the temperate zones, can be easily cultivated. Java too possesses a civilization, a history and antiquities of its own, of great interest. The Brahminical religion flourished in it from an epoch of unknown antiquity until about the year 1478, when that of Mahomet superseded it. The former religion was accompanied by a civilization which has not been equalled by the conquerors; for, scattered through the country, especially in the eastern part of it, are found buried in lofty forests, temples, tombs, and statues of great beauty and grandeur; and the remains of extensive cities, where the tiger, the rhinoceros, and the wild bull now roam undisturbed. A modern civilization of another type is now spreading over the land. Good roads run through the country from end to end; European and native rulers work harmoniously together; and life and property are as well secured as in the best governed states of Europe. I believe, therefore, that Java may fairly claim to be the finest tropical island in the world, and equally interesting to the tourist seeking after new and beautiful scenes; to the naturalist who desires to examine the variety and beauty of tropical nature; or to the moralist and the politician who want to solve the problem of how man may be best governed under new and varied conditions.

Volume I

The Dutch mail steamer brought me from Ternate to Sourabaya, the chief town and port in the eastern part of Java, and after a fortnight spent in packing up and sending off my last collections, I started on a short journey into the interior. Travelling in Java is very luxurious but very expensive, the only way being to hire or borrow a carriage, and then pay half a crown a mile for post-horses, which are changed at regular posts every six miles, and will carry you at the rate of ten miles an hour from one end of the island to the other. Bullock carts or coolies are required to carry all extra baggage. As this kind of travelling would not suit my means, I determined on making only a short journey to the district at the foot of Mount Arjuna, where I was told there were extensive forests, and where I hoped to be able to make some good collections. The country for many miles behind Sourabaya is perfectly flat and everywhere cultivated; being a delta or alluvial plain, watered by many branching streams. Immediately around the town the evident signs of wealth and of an industrious population were very pleasing; but as we went on, the constant succession of open fields skirted by rows of bamboos, with here and there the white buildings and a tall chimney of a sugar-mill, became monotonous. The roads run in straight lines for several miles at a stretch, and are bordered by rows of dusty tamarind-trees. At each mile there are little guardhouses, where a policeman is stationed; and there is a wooden gong, which by means of concerted signals may be made to convey information over the country with great rapidity. About every six or seven miles is the posthouse, where the horses are changed as quickly as were those of the mail in the old coaching days in England.

I stopped at Modjo-kerto, a small town about forty miles south of Sourabaya, and the nearest point on the high road to the district I wished to visit. I had a letter of introduction to Mr. Ball, an Englishman, long resident in Java and married to a Dutch lady; and he kindly invited me to stay with him until I could fix on a place to suit me. A Dutch Assistant Resident as well as a Regent or native Javanese prince lived here. The town was neat, and had a nice open grassy space like a village green, on which stood a magnificent fig-tree (allied to the Banyan of India, but more lofty), under whose shade a kind of market is continually held, and where the inhabitants meet together to lounge and chat. The day after my arrival, Mr. Ball drove me over to the village of Modjo-agong, where he was building a house and premises for the tobacco trade, which is carried on here by a system of native cultivation and advance purchase, somewhat similar to the indigo trade in British India. On our way we stayed to look at a fragment of the ruins of the ancient city of Modjo-pahit, consisting of two lofty brick masses, apparently the sides of a gateway. The extreme perfection and beauty of the brickwork astonished me. The bricks are exceedingly fine and hard, with sharp angles and true surfaces. They are laid with great exactness, without visible mortar or cement, yet somehow fastened together so that the joints are hardly perceptible, and sometimes the two surfaces coalesce in a most incomprehensible manner.

Such admirable brickwork I have never seen before or since. There was no sculpture here, but an abundance of bold projections and finely-worked mouldings. Traces of buildings exist for many miles in every direction, and almost every road and pathway shows a foundation of brickwork beneath it—the paved roads of the old city. In the house of the Waidono or district chief at Modjo-agong, I saw a beautiful figure carved in high relief out of a block of lava, and which had been found buried in the ground near the village. On my expressing a wish to obtain some such specimen, Mr. B. asked the chief for it, and much to my surprise he immediately gave it me. It represented the Hindu goddess Durga, called in Java, Lora Jong-grang (the exalted virgin). She has eight arms, and stands on the back of a kneeling bull. Her lower right hand holds the tail of the bull, while the corresponding left hand grasps the hair of a captive, Dewth Mahikusor, the personification of vice, who has attempted to slay her bull. He has a cord round his waist, and crouches at her feet in an attitude of supplication. The other hands of the goddess hold, on her right side, a double hook or small anchor, a broad straight sword, and a noose of thick cord; on her left, a girdle or armlet of large beads or shells, an unstrung bow, and a standard or war flag. This deity was a special favourite among the old Javanese, and her image is often found in the ruined temples which abound in the eastern part of the island.

The specimen I had obtained was a small one, about two feet high, weighing perhaps a hundredweight; and the next day we had it conveyed to Modjo-Kerto to await my return to Sourabaya. Having decided to stay some time at Wonosalem, on the lower slopes of the Arjuna Mountain, where I was informed I should find forest and plenty of game, I had first to obtain a recommendation from the Assistant Resident to the Regent, and then an order from the Regent to the Waidono; and when after a week's delay I arrived with my baggage and men at Modjo-agong, I found them all in the midst of a five days' feast, to celebrate the circumcision of the Waidono's younger brother and cousin, and had a small room in an on outhouse given me to stay in. The courtyard and the great open reception-shed were full of natives coming and going and making preparations for a feast which was to take place at midnight, to which I was invited, but preferred going to bed. A native band, or Gamelang, was playing almost all the evening, and I had a good opportunity of seeing the instruments and musicians. The former are chiefly gongs of various sizes, arranged in sets of from eight to twelve, on low wooden frames. Each set is played by one performer with one or two drumsticks.

Volume I

There are also some very large gongs, played singly or in pairs, and taking the place of our drums and kettledrums. Other instruments are formed by broad metallic bars, supported on strings stretched across frames; and others again of strips of bamboo similarly placed and producing the highest notes. Besides these there were a flute and a curious two-stringed violin, requiring in all twenty-four performers. There was a conductor, who led off and regulated the time, and each performer took his part, coming in occasionally with a few bars so as to form a harmonious combination. The pieces played were long and complicated, and some of the players were mere boys, who took their parts with great precision. The general effect was very pleasing, but, owing to the similarity of most of the instruments, more like a gigantic musical box than one of our bands; and in order to enjoy it thoroughly it is necessary to watch the large number of performers who are engaged in it. The next morning, while I was waiting for the men and horses who were to take me and my baggage to my destination, the two lads, who were about fourteen years old, were brought out, clothed in a sarong from the waist downwards, and having the whole body covered with yellow powder, and profusely decked with white blossom in wreaths, necklaces, and armlets, looking at first sight very like savage brides. They were conducted by two priests to a bench placed in front of the house in the open air, and the ceremony of circumcision was then performed before the assembled crowd.

The road to Wonosalem led through a magnificent forest in the depths of which we passed a fine ruin of what appeared to have been a royal tomb or mausoleum. It is formed entirely of stone, and elaborately carved. Near the base is a course of boldly projecting blocks, sculptured in high relief, with a series of scenes which are probably incidents in the life of the defunct. These are all beautifully executed, some of the figures of animals in particular, being easily recognisable and very accurate. The general design, as far as the ruined state of the upper part will permit of its being seen, is very good, effect being given by an immense number and variety of projecting or retreating courses of squared stones in place of mouldings. The size of this structure is about thirty feet square by twenty high, and as the traveller comes suddenly upon it on a small elevation by the roadside, overshadowed by gigantic trees, overrun with plants and creepers, and closely backed by the gloomy forest, he is struck by the solemnity and picturesque beauty of the scene, and is led to ponder on the strange law of progress, which looks so like retrogression, and which in so many distant parts of the world has exterminated or driven out a highly artistic and constructive race, to make room for one which, as far as we can judge, is very far its inferior.

Few Englishmen are aware of the number and beauty of the architectural remains in Java. They have never been popularly illustrated or described, and it will therefore take most persons by surprise to learn that they far surpass those of Central America, perhaps even those of India. To give some idea of these ruins, and perchance to excite wealthy amateurs to explore them thoroughly and obtain by photography an accurate record of their beautiful sculptures before it is too late, I will enumerate the most important, as briefly described in Sir Stamford Raffles' "History of Java."

BRAMBANAM.—Near the centre of Java, between the native capitals of Djoko-kerta and Surakerta, is the village of Brambanam, near which are abundance of ruins, the most important being the temples of Loro-Jongran and Chandi Sewa. At Loro-Jongran there were twenty separate buildings, six large and fourteen small temples. They are now a mass of ruins, but the largest temples are supposed to have been ninety feet high. They were all constructed of solid stone, everywhere decorated with carvings and bas-reliefs, and adorned with numbers of statues, many of which still remain entire. At Chandi Sewa, or the "Thousand Temples," are many fine colossal figures. Captain Baker, who surveyed these ruins, said he had never in his life seen "such stupendous and finished specimens of human labour, and of the science and taste of ages long since forgot, crowded together in so small a compass as in this spot." They cover a space of nearly six hundred feet square, and consist of an outer row of eighty-four small temples, a second row of seventy-six, a third of sixty-four, a fourth of forty-four, and the fifth forming an inner parallelogram of twenty-eight, in all two hundred and ninety-six small temples; disposed in five regular parallelograms. In the centre is a large cruciform temple surrounded by lofty flights of steps richly ornamented with sculpture, and containing many apartments. The tropical vegetation has ruined most of the smaller temples, but some remain tolerably perfect, from which the effect of the whole may be imagined.

About half a mile off is another temple, called Chandi Kali Bening, seventy-two feet square and sixty feet high, in very fine preservation, and covered with sculptures of Hindu mythology surpassing any that exist in India, other ruins of palaces, halls, and temples, with abundance of sculptured deities, are found in the same neighbourhood.

BOROBODO.—About eighty miles westward, in the province of Kedu, is the great temple of Borobodo. It is built upon a small hill, and consists of a central dome and seven ranges of terraced walls covering the slope of the hill and forming open galleries each below the other, and communicating by steps and gateways. The central dome is fifty feet in diameter; around it is a triple circle of seventy-two towers, and the whole building is six hundred and twenty feet square, and about one hundred feet high. In the terrace walls are niches containing cross-legged figures larger than life to the number of about four hundred, and both sides of all the terrace walls are covered with bas-reliefs crowded with figures, and carved in hard stone and which must therefore occupy an extent of nearly three miles in length! The amount of human labour and skill expended on the Great Pyramid of Egypt sinks into insignificance when compared with that required to complete this sculptured hill-temple in the interior of Java.

GUNONG PRAU.—About forty miles southwest of Samarang, on a mountain called Gunong Prau, an extensive plateau is covered with ruins. To reach these temples, four flights of stone steps were made up the mountain from opposite directions, each flight consisting of more than a thousand steps. Traces of nearly four hundred temples have been found here, and many (perhaps all) were decorated with rich and delicate sculptures. The whole country between this and Brambanam, a distance of sixty miles, abounds with ruins, so that fine sculptured images may be seen lying in the ditches, or built into the walls of enclosures.

In the eastern part of Java, at Kediri and in Malang, there are equally abundant traces of antiquity, but the buildings themselves have been mostly destroyed. Sculptured figures, however, abound; and the ruins of forts, palaces, baths, aqueducts, and temples, can be everywhere traced. It is altogether contrary to the plan of this book to describe what I have not myself seen; but, having been led to mention them, I felt bound to do something to call attention to these marvellous works of art. One is overwhelmed by the contemplation of these innumerable sculptures, worked with delicacy and artistic feeling in a hard, intractable, trachytic rock, and all found in one tropical island. What could have been the state of society, what the amount of population, what the means of subsistence which rendered such gigantic works possible, will, perhaps, ever remain a mystery; and it is a wonderful example of the power of religious ideas in social life, that in the very country where, five hundred years ago, these grand works were being yearly executed, the inhabitants now only build rude houses of bamboo and thatch, and look upon these relics of their forefathers with ignorant amazement, as the undoubted productions of giants or of demons. It is much to be regretted that the Dutch Government does not take vigorous steps for the preservation of these ruins from the destroying agency of tropical vegetation; and for the collection of the fine sculptures which are everywhere scattered over the land.

Wonosalem is situated about a thousand feet above the sea, but unfortunately it is at a distance from the forest, and is surrounded by coffee plantations, thickets of bamboo, and coarse grasses. It was too far to walk back daily to the forest, and in other directions I could find no collecting ground for insects. The place was, however, famous for peacocks, and my boy soon shot several of these magnificent birds, whose flesh we found to be tender, white, and delicate, and similar to that of a turkey. The Java peacock is a different species

Volume I

from that of India, the neck being covered with scale-like green feathers, and the crest of a different form; but the eyed train is equally large and equally beautiful. It is a singular fact in geographical distribution that the peacock should not be found in Sumatra or Borneo, while the superb Argus, Fire-backed and Ocellated pheasants of those islands are equally unknown in Java. Exactly parallel is the fact that in Ceylon and Southern India, where the peacock abounds, there are none of the splendid Lophophori and other gorgeous pheasants which inhabit Northern India. It would seem as if the peacock can admit of no rivals in its domain. Were these birds rare in their native country, and unknown alive in Europe, they would assuredly be considered as the true princes of the feathered tribes, and altogether unrivalled for stateliness and beauty. As it is, I suppose scarcely anyone if asked to fix upon the most beautiful bird in the world would name the peacock, any more than the Papuan savage or the Bugis trader would fix upon the bird of paradise for the same honour.

Three days after my arrival at Wonosalem, my friend Mr. Ball came to pay me a visit. He told me that two evenings before, a boy had been killed and eaten by a tiger close to Modjo-agong. He was riding on a cart drawn by bullocks, and was coming home about dusk on the main road; and when not half a mile from the village a tiger sprang upon him, carried him off into the jungle close by, and devoured him. Next morning his remains were discovered, consisting only of a few mangled bones. The Waidono had got together about seven hundred men, and were in chase of the animal, which, I afterwards heard, they found and killed. They only use spears when in pursuit of a tiger in this way. They surround a large tract of country, and draw gradually together until the animal is enclosed in a compact ring of armed men. When he sees there is no escape he generally makes a spring, and is received on a dozen spears, and almost instantly stabbed to death. The skin of an animal thus killed is, of course, worthless, and in this case the skull, which I had begged Mr. Ball to secure for me, was hacked to pieces to divide the teeth, which are worn as charms.

After a week at Wonosalem, I returned to the foot of the mountain, to a village named Djapannan, which was surrounded by several patches of forest, and seemed altogether pretty well suited to my pursuits. The chief of the village had prepared two small bamboo rooms on one side of his own courtyard to accommodate me, and seemed inclined to assist me as much as he could. The weather was exceedingly hot and dry, no rain having fallen for several months, and there was, in consequence, a great scarcity of insects, and especially of beetles. I therefore devoted myself chiefly to obtaining a good set of the birds, and succeeded in making a tolerable collection. All the peacocks we had hitherto shot had had short or imperfect tails, but I now obtained two magnificent specimens more than seven feet long, one of which I preserved entire, while I kept the train only attached to the tail of two or three others. When this bird is seen feeding on the ground, it appears wonderful how it can rise into the air with such a long and cumbersome train of feathers. It does so however with great ease, by running quickly for a short distance, and then rising obliquely; and will fly over trees of a considerable height. I also obtained here a specimen of the rare green jungle-fowl (Gallus furcatus), whose back and neck are beautifully scaled with bronzy feathers, and whose smooth-edged oval comb is of a violet purple colour, changing to green at the base. It is also remarkable in possessing a single large wattle beneath its throat, brightly coloured in three patches of red, yellow, and blue. The common jungle-cock (Gallus bankiva) was also obtained here. It is almost exactly like a common game-cock, but the voice is different, being much shorter and more abrupt; hence its native name is Bekeko. Six different kinds of woodpeckers and four kingfishers were found here, the fine hornbill, Buceros lunatus, more than four feet long, and the pretty little lorikeet, Loriculus pusillus, scarcely more than as many inches.

One morning, as I was preparing and arranging specimens, I was told there was to be a trial; and presently four or five men came in and squatted down on a mat under the audience-shed in the court. The chief then came in with his clerk, and sat down opposite them. Each spoke in turn, telling his own tale, and then I found that those who first entered were the prisoner, accuser, policemen, and witness, and that the prisoner was indicated solely by having a loose piece of cord twined around his wrists, but not tied. It was a case of robbery, and after the evidence was given, and a few questions had been asked by the chief, the accused said a few words, and then sentence was pronounced, which was a fine. The parties then got up and walked away together, seeming quite friendly; and throughout there was nothing in the manner of any one present indicating passion or ill-feeling—a very good illustration of the Malayan type of character.

In a month's collecting at Wonosalem and Djapannan I accumulated ninety-eight species of birds, but a most miserable lot of insects. I then determined to leave East Java and try the more moist and luxuriant districts at the western extremity of the island. I returned to Sourabaya by water, in a roomy boat which brought myself, servants, and baggage at one-fifth the expense it had cost me to come to Modjo-kerto. The river has been rendered navigable by being carefully banked up, but with the usual effect of rendering the adjacent country liable occasionally to severe floods. An immense traffic passes down this river; and at a lock we passed through, a mile of laden boats were waiting two or three deep, which pass through in their turn six at a time.

Volume I

A few days afterwards I went by steamer to Batavia, where I stayed about a week at the chief hotel, while I made arrangements for a trip into the interior. The business part of the city is near the harbour, but the hotels and all the residences of the officials and European merchants are in a suburb two miles off, laid out in wide streets and squares so as to cover a great extent of ground. This is very inconvenient for visitors, as the only public conveyances are handsome two-horse carriages, whose lowest charge is five guilders (8s. 4d.) for half a day, so that an hour's business in the morning and a visit in the evening costs 16s. 8d. a day for carriage hire alone.

Batavia agrees very well with Mr. Money's graphic account of it, except that his "clear canals" were all muddy, and his "smooth gravel drives" up to the houses were one and all formed of coarse pebbles, very painful to walk upon, and hardly explained by the fact that in Batavia everybody drives, as it can hardly be supposed that people never walk in their gardens. The Hôtel des Indes was very comfortable, each visitor having a sitting-room and bedroom opening on a verandah, where he can take his morning coffee and afternoon tea. In the centre of the quadrangle is a building containing a number of marble baths always ready for use; and there is an excellent table d'hôte breakfast at ten, and dinner at six, for all which there is a moderate charge per day.

I went by coach to Buitenzorg, forty miles inland and about a thousand feet above the sea, celebrated for its delicious climate and its Botanical Gardens. With the latter I was somewhat disappointed. The walks were all of loose pebbles, making any lengthened wanderings about them very tiring and painful under a tropical sun. The gardens are no doubt wonderfully rich in tropical and especially in Malayan plants, but there is a great absence of skillful laying-out; there are not enough men to keep the place thoroughly in order, and the plants themselves are seldom to be compared for luxuriance and beauty to the same species grown in our hothouses. This can easily be explained. The plants can rarely be placed in natural or very favourable conditions. The climate is either too hot or too cool, too moist or too dry, for a large proportion of them, and they seldom get the exact quantity of shade or the right quality of soil to suit them. In our stoves these varied conditions can be supplied to each individual plant far better than in a large garden, where the fact that the plants are most of them growing in or near their native country is supposed to preclude, the necessity of giving them much individual attention. Still, however, there is much to admire here. There are avenues of stately palms, and clumps of bamboos of perhaps fifty different kinds; and an endless variety of tropical shrubs and trees with strange and beautiful foliage. As a change from the excessive heat of Batavia, Buitenzorg is a delightful abode.

It is just elevated enough to have deliciously cool evenings and nights, but not so much as to require any change of clothing; and to a person long resident in the hotter climate of the plains, the air is always fresh and pleasant, and admits of walking at almost any hour of the day. The vicinity is most picturesque and luxuriant, and the great volcano of Gunung Salak, with its truncated and jagged summit, forms a characteristic background to many of the landscapes. A great mud eruption took place in 1699, since which date the mountain has been entirely inactive.

On leaving Buitenzorg, I had coolies to carry my baggage and a horse for myself, both to be changed every six or seven miles. The road rose gradually, and after the first stage the hills closed in a little on each side, forming a broad valley; and the temperature was so cool and agreeable, and the country so interesting, that I preferred walking. Native villages imbedded in fruit trees, and pretty villas inhabited by planters or retired Dutch officials, gave this district a very pleasing and civilized aspect; but what most attracted my attention was the system of terrace-cultivation, which is here universally adopted, and which is, I should think, hardly equalled in the world. The slopes of the main valley, and of its branches, were everywhere cut in terraces up to a considerable height, and when they wound round the recesses of the hills produced all the effect of magnificent amphitheatres. Hundreds of square miles of country are thus terraced, and convey a striking idea of the industry of the people and the antiquity of their civilization. These terraces are extended year by year as the population increases, by the inhabitants of each village working in concert under the direction of their chiefs; and it is perhaps by this system of village culture alone, that such extensive terracing and irrigation has been rendered possible. It was probably introduced by the Brahmins from India, since in those Malay countries where there is no trace of a previous occupation by a civilized people, the terrace system is unknown. I first saw this mode of cultivation in Bali and Lombock, and, as I shall have to describe it in some detail there (see CHAPTER X.), I need say no more about it in this place, except that, owing to the finer outlines and greater luxuriance of the country in West Java, it produces there the most striking and picturesque effect. The lower slopes of the mountains in Java possess such a delightful climate and luxuriant soil; living is so cheap and life and property are so secure, that a considerable number of Europeans who have been engaged in Government service, settle permanently in the country instead of returning to Europe. They are scattered everywhere throughout the more accessible parts of the island, and tend greatly to the gradual improvement of the native population, and to the continued peace and prosperity of the whole country.

Twenty miles beyond Buitenzorg the post road passes over the Megamendong Mountain, at an elevation of about

4,500 feet. The country is finely mountainous, and there is much virgin forest still left upon the hills, together with some of the oldest coffee-plantations in Java, where the plants have attained almost the dimensions of forest trees. About 500 feet below the summit level of the pass there is a road-keeper's hut, half of which I hired for a fortnight, as the country looked promising for making collections. I almost immediately found that the productions of West Java were remarkably different from those of the eastern part of the island; and that all the more remarkable and characteristic Javanese birds and insects were to be found here. On the very first day, my hunters obtained for me the elegant yellow and green trogon (Harpactes Reinwardti), the gorgeous little minivet flycatcher (Pericrocotus miniatus), which looks like a flame of fire as it flutters among the bushes, and the rare and curious black and crimson oriole (Analcipus sanguinolentus), all of these species which are found only in Java, and even seem to be confined to its western portion.

In a week I obtained no less than twenty-four species of birds, which I had not found in the east of the island, and in a fortnight this number increased to forty species, almost all of which are peculiar to the Javanese fauna. Large and handsome butterflies were also tolerably abundant. In dark ravines, and occasionally on the roadside, I captured the superb Papilio arjuna, whose wings seem powdered with grains of golden green, condensed into bands and moon-shaped spots; while the elegantly-formed Papilio coön was sometimes to be found fluttering slowly along the shady pathways (see figure at page 201). One day a boy brought me a butterfly between his fingers, perfectly unhurt. He had caught it as it was sitting with wings erect, sucking up the liquid from a muddy spot by the roadside. Many of the finest tropical butterflies have this habit, and they are generally so intent upon their meal that they can be easily be reached and captured. It proved to be the rare and curious Charaxes kadenii, remarkable for having on each hind wing two curved tails like a pair of callipers. It was the only specimen I ever saw, and is still the only representative of its kind in English collections.

In the east of Java I had suffered from the intense heat and drought of the dry season, which had been very inimical to insect life. Here I had got into the other extreme of damp, wet, and cloudy weather, which was equally unfavourable. During the month which I spent in the interior of West Java, I never had a really hot fine day throughout. It rained almost every afternoon, or dense mists came down from the mountains, which equally stopped collecting, and rendered it most difficult to dry my specimens, so that I really had no chance of getting a fair sample of Javanese entomology.

By far the most interesting incident in my visit to Java was a trip to the summit of the Pangerango and Gedeh mountains; the former an extinct volcanic cone about 10,000 feet high, the latter an active crater on a lower portion of the same mountain range. Tchipanas, about four miles over the Megamendong Pass, is at the foot of the mountain. A small country house for the Governor-General and a branch of the Botanic Gardens are situated here, the keeper of which accommodated me with a bed for a night. There are many beautiful trees and shrubs planted here, and large quantities of European vegetables are grown for the Governor-General's table. By the side of a little torrent that bordered the garden, quantities of orchids were cultivated, attached to the trunks of trees, or suspended from the branches, forming an interesting open air orchid-house. As I intended to stay two or three nights on the mountain, I engaged two coolies to carry my baggage, and with my two hunters we started early the next morning.

The first mile was over open country, which brought us to the forest that covers the whole mountain from a height of about 5,000 feet. The next mile or two was a tolerably steep ascent through a grand virgin forest, the trees being of great size, and the undergrowth consisting of fine herbaceous plants, tree-ferns, and shrubby vegetation. I was struck by the immense number of ferns that grew by the side of the road. Their variety seemed endless, and I was continually stopping to admire some new and interesting forms. I could now well understand what I had been told by the gardener, that 300 species had been found on this one mountain. A little before noon we reached the small plateau of Tjiburong, at the foot of the steeper part of the mountain, where there is a plank-house for the accommodation of travellers. Close by is a picturesque waterfall and a curious cavern, which I had not time to explore. Continuing our ascent the road became narrow, rugged and steep, winding zigzag up the cone, which is covered with irregular masses of rock, and overgrown with a dense luxuriant but less lofty vegetation. We passed a torrent of water which is not much lower than the boiling point, and has a most singular appearance as it foams over its rugged bed, sending up clouds of steam, and often concealed by the overhanging herbage of ferns and lycopodia, which here thrive with more luxuriance than elsewhere.

At about 7,500 feet we came to another hut of open bamboos, at a place called Kandang Badak, or "Rhinoceros-field," which we were going to make our temporary abode. Here was a small clearing, with abundance of tree-ferns and some young plantations of Cinchona. As there was now a thick mist and drizzling rain, I did not attempt to go on to the summit that evening, but made two visits to it during my stay, as well as one to the active crater of Gedeh. This is a vast semicircular chasm, bounded by black perpendicular walls of rock, and surrounded by miles of rugged scoria-covered slopes. The crater itself is not very deep. It exhibits patches of sulphur and variously-coloured volcanic products, and emits from several vents continual streams of smoke and vapour. The extinct cone of Pangerango was to me more interesting. The summit is an irregular undulating plain with a low bordering ridge, and one deep lateral chasm. Unfortunately, there was perpetual mist and rain either above or below us all the time I was on the mountain; so that I never once saw the plain below, or had a glimpse of the magnificent view which in fine weather is to be obtained from its summit. Notwithstanding this drawback I enjoyed the excursion exceedingly, for it was the first time I had been high enough on a mountain near the Equator to watch the change from a tropical to a temperate flora. I will now briefly sketch these changes as I observed them in Java.

On ascending the mountain, we first meet with temperate forms of herbaceous plants, so low as 3,000 feet, where strawberries and violets begin to grow, but the former are tasteless, and the latter have very small and pale flowers. Weedy composites also begin to give a European aspect to the wayside herbage. It is between 2,000 and 5,000 feet that the forests and ravines exhibit the utmost development of tropical luxuriance and beauty. The abundance of noble Tree-ferns, sometimes fifty feet high, contributes greatly to the general effect, since of all the forms of tropical vegetation they are certainly the most striking and beautiful. Some of the deep ravines which have been cleared of large timber are full of them from top to bottom; and where the road crosses one of these valleys, the view of their feathery crowns, in varied positions above and below the eye, offers a spectacle of picturesque beauty never to be forgotten. The splendid foliage of the broad-leaved Musaceae and Zingiberaceae, with their curious and brilliant flowers; and the elegant and varied forms of plants allied to Begonia and Melastoma, continually attract the attention in this region. Filling in the spaces between the trees and larger plants, on every trunk and stump and branch, are hosts of Orchids, Ferns and Lycopods, which wave and hang and intertwine in ever-varying complexity. At about 5,000 feet I first saw horsetails (Equisetum), very like our own species. At 6,000 feet, raspberries abound, and thence to the summit of the mountain there are three species of eatable Rubus. At 7,000 feet Cypresses appear, and the forest trees become reduced in size, and more covered with mosses and lichens. From this point upward these rapidly increase, so that the blocks of rock and scoria that form the mountain slope are completely hidden in a mossy vegetation. At about 5,000 feet European forms of plants become abundant. Several species of Honeysuckle, St. John's-wort, and Guelder-rose abound, and at about 9,000 feet we first meet with the rare and beautiful Royal Cowslip (Primula imperialis), which is said to be found nowhere else in the world but on this solitary mountain summit. It has a tall, stout stem, sometimes more than three feet high, the root leaves are eighteen inches long, and it bears several whorls of cowslip-like flowers, instead of a terminal cluster only. The forest trees, gnarled and dwarfed to the dimensions of bushes, reach up to the very rim of the old crater, but do not extend over the hollow on its summit. Here we find a good deal of open ground, with thickets of shrubby Artemisias and Gnaphaliums, like our southernwood and cudweed, but six or eight feet high; while Buttercups, Violets, Whortleberries, Sow-thistles, Chickweed, white and yellow Cruciferae, Plantain, and annual grasses everywhere abound. Where there are bushes and shrubs, the St. John's-wort and Honeysuckle grow abundantly, while the Imperial Cowslip only exhibits its elegant blossoms under the damp shade of the thickets.

Mr. Motley, who visited the mountain in the dry season, and paid much attention to botany, gives the following list of genera of European plants found on or near the summit:— Two species of Violet, three of Ranunculus, three of Impatiens, eight or ten of Rubus, and species of Primula, Hypericum, Swertia, Convallaria (Lily of the Valley), Vaccinium (Cranberry), Rhododendron, Gnaphalium, Polygonum, Digitalis (Foxglove), Lonicera (Honeysuckle), Plantago (Rib-grass), Artemisia (Wormwood), Lobelia, Oxalis (Wood-sorrel), Quercus (Oak), and Taxus (Yew). A few of the smaller plants (Plantago major and lanceolata, Sonchus oleraceus, and Artemisia vulgaris) are identical with European species.

The fact of a vegetation so closely allied to that of Europe occurring on isolated mountain peaks, in an island south of the Equator, while all the lowlands for thousands of miles around are occupied by a flora of a totally different character, is very extraordinary; and has only recently received an intelligible explanation. The Peak of Teneriffe, which rises to a greater height and is much nearer to Europe, contains no such Alpine flora; neither do the mountains of Bourbon and Mauritius. The case of the volcanic peaks of Java is therefore somewhat exceptional, but there are several analogous, if not exactly parallel cases, that will enable us better to understand in what way the phenomena may possibly have been brought about.

The higher peaks of the Alps, and even of the Pyrenees, contain a number of plants absolutely identical with those of Lapland, but nowhere found in the intervening plains. On the summit of the White Mountains, in the United States, every plant is identical with species growing in Labrador. In these cases all ordinary means of transport fail. Most of the plants have heavy seeds, which could not possibly be carried such immense distances by the wind; and the agency of birds in so effectually stocking these Alpine heights is equally out of the question. The difficulty was so great, that some naturalists were driven to believe that these species were all separately created twice over on these distant peaks. The determination of a recent glacial epoch, however, soon offered a much more satisfactory solution, and one that is now universally accepted by men of science. At this period, when the mountains of Wales were full of glaciers, and the mountainous parts of Central Europe, and much of America north of the great lakes, were covered with snow and ice, and had a climate resembling that of Labrador and Greenland at the present day, an Arctic flora covered all these regions. As this epoch of cold passed away, and the snowy mantle of the country, with the glaciers that descended from every mountain summit, receded up their slopes and towards the north pole, the plants receded also, always clinging as now to the margins of the perpetual snow line. Thus it is that the same species are now found on the summits of the mountains of temperate Europe and America, and in the barren north-polar regions.

But there is another set of facts, which help us on another step towards the case of the Javanese mountain flora. On the higher slopes of the Himalayas, on the tops of the mountains of Central India and of Abyssinia, a number of plants occur which, though not identical with those of European mountains, belong to the same genera, and are said by botanists to represent them; and most of these could not exist in the warm intervening plains. Mr. Darwin believes that this class of facts can be explained in the same way; for, during the greatest severity of the glacial epoch, temperate forms of plants will have extended to the confines of the tropics, and on its departure, will have retreated up these southern mountains, as well as northward to the plains and hills of Europe. But in this case, the time elapsed, and the great change of conditions, have allowed many of these plants to become so modified that we now consider them to be distinct species. A variety of other facts of a similar nature have led him to believe that the depression of temperature was at one time sufficient to allow a few north-temperate plants to cross the Equator (by the most elevated routes) and to reach the Antarctic regions, where they are now found. The evidence on which this belief rests will be found in the latter part of CHAPTER II. of the "Origin of Species"; and, accepting it for the present as an hypothesis, it enables us to account for the presence of a flora of European type on the volcanoes of Java.

It will, however, naturally be objected that there is a wide expanse of sea between Java and the continent, which would have effectually prevented the immigration of temperate forms of plants during the glacial epoch. This would undoubtedly be a fatal objection, were there not abundant evidence to show that Java has been formerly connected with Asia, and that the union must have occurred at about the epoch required. The most striking proof of such a junction is, that the great Mammalia of Java, the rhinoceros, the tiger, and the Banteng or wild ox, occur also in Siam and Burmah, and these would certainly not have been introduced by man. The Javanese peacock and several other birds are also common to these two countries; but, in the majority of cases, the species are distinct, though closely allied, indicating that a considerable time (required for such modification) has elapsed since the separation, while it has not been so long as to cause an entire change. Now this exactly corresponds with the time we should require since the temperate forms of plants entered Java. These are now almost distinct species, but the changed conditions under which they are now forced to exist, and the probability of some of them having since died out on the continent of India, sufficiently accounts for the Javanese species being different.

In my more special pursuits, I had very little success upon the mountain—owing, perhaps, to the excessively unpropitious weather and the shortness of my stay. At from 7,000 to 8,000 feet elevation, I obtained one of the most lovely of the small Fruit pigeons (Ptilonopus roseicollis), whose entire head and neck are of an exquisite rosy pink colour, contrasting finely with its otherwise green plumage; and on the very summit, feeding on the ground among the strawberries that have been planted there, I obtained a dull-coloured thrush, with the form and habits of a starling (Turdus fumidus). Insects were almost entirely absent, owing no doubt to the extreme dampness, and I did not get a single butterfly the whole trip; yet I feel sure that, during the dry season, a week's residence on this mountain would well repay the collector in every department of natural history.

After my return to Toego, I endeavoured to find another locality to collect in, and removed to a coffee-plantation some miles to the north, and tried in succession higher and lower stations on the mountain; but, I never succeeded in obtaining insects in any abundance and birds were far less plentiful than on the Megamendong Mountain. The weather now became more rainy than ever, and as the wet season seemed to have set in in earnest, I returned to Batavia, packed up and sent off my collections, and left by steamer on November 1st for Banca and Sumatra.

CHAPTER VIII. SUMATRA.

(NOVEMBER 1861 to JANUARY 1862.)

The mail steamer from Batavia to Singapore took me to Muntok (or as on English maps, "Minto"), the chief town and port of Banca. Here I stayed a day or two, until I could obtain a boat to take me across the straits, and up the river to Palembang. A few walks into the country showed me that it was very hilly, and full of granitic and laterite rocks, with a dry and stunted forest vegetation; and I could find very few insects. A good-sized open sailing-boat took me across to the mouth of the Palembang river where, at a fishing village, a rowing-boat was hired to take me up to Palembang—a distance of nearly a hundred miles by water. Except when the wind was strong and favourable we could only proceed with the tide, and the banks of the river were generally flooded Nipa-swamps, so that the hours we were obliged to lay at anchor passed very heavily. Reaching Palembang on the 8th of November, I was lodged by the Doctor, to whom I had brought a letter of introduction, and endeavoured to ascertain where I could find a good locality for collecting. Everyone assured me that I should have to go a very long way further to find any dry forest, for at this season the whole country for many miles inland was flooded. I therefore had to stay a week at Palembang before I could determine my future movements.

The city is a large one, extending for three or four miles along a fine curve of the river, which is as wide as the Thames at Greenwich. The stream is, however, much narrowed by the houses which project into it upon piles, and within these, again, there is a row of houses built upon great bamboo rafts, which are moored by rattan cables to the shore or to piles, and rise and fall with the tide.

The whole riverfront on both sides is chiefly formed of such houses, and they are mostly shops open to the water, and only raised a foot above it, so that by taking a small boat it is easy to go to market and purchase anything that is to be had in Palembang. The natives are true Malays, never building a house on dry land if they can find water to set it in, and never going anywhere on foot if they can reach the place in a boat. A considerable portion of the population are Chinese and Arabs, who carry on all the trade; while the only Europeans are the civil and military officials of the Dutch Government. The town is situated at the head of the delta of the river, and between it and the sea there is very little ground elevated above highwater mark; while for many miles further inland, the banks of the main stream and its numerous tributaries are swampy, and in the wet season flooded for a considerable distance. Palembang is built on a patch of elevated ground, a few miles in extent, on the north bank of the river. At a spot about three miles from the town this turns into a little hill, the top of which is held sacred by the natives, shaded by some fine trees, and inhabited by a colony of squirrels which have become half-tame. On holding out a few crumbs of bread or any fruit, they come running down the trunk, take the morsel out of your fingers, and dart away instantly. Their tails are carried erect, and the hair, which is ringed with grey, yellow, and brown, radiates uniformly around them, and looks exceedingly pretty. They have somewhat of the motions of mice, coming on with little starts, and gazing intently with their large black eyes before venturing to advance further. The manner in which Malays often obtain the confidence of wild animals is a very pleasing trait in their character, and is due in some degree to the quiet deliberation of their manners, and their love of repose rather than of action. The young are obedient to the wishes of their elders, and seem to feel none of that propensity to mischief which European boys exhibit. How long would tame squirrels continue to inhabit trees in the vicinity of an English village, even if close to the church? They would soon be pelted and driven away, or snared and confined in a whirling cage. I have never heard of these pretty animals being tamed in this way in England, but I should think it might be easily done in any gentleman's park, and they would certainly be as pleasing and attractive as they would be uncommon.

After many inquiries, I found that a day's journey by water above Palembang there commenced a military road which extended up to the mountains and even across to Bencoolen, and I determined to take this route and travel on until I found some tolerable collecting ground. By this means I should secure dry land and a good road, and avoid the rivers, which at this season are very tedious to ascend owing to the powerful currents, and very unproductive to the collector owing to most of the lands in their vicinity being underwater. Leaving early in the morning we did not reach Lorok, the village where the road begins, until late at night. I stayed there a few days, but found that almost all the ground in the vicinity not underwater was cultivated, and that the only forest was in swamps which were now inaccessible. The only bird new to me which I obtained at Lorok was the fine long-tailed parroquet (Palaeornis longicauda). The people here assured me that the country was just the same as this for a very long way—more than a week's journey, and they seemed hardly to have any conception of an elevated forest-clad country, so that I began to think it would be useless going on, as the time at my disposal was too short to make it worth my while to spend much more of it in moving about. At length, however, I found a man who knew the country, and was more intelligent; and he at once told me that if I wanted forest I must go to the district of Rembang, which I found on inquiry was about twenty-five or thirty miles off.

The road is divided into regular stages of ten or twelve miles each, and, without sending on in advance to have coolies ready, only this distance can be travelled in a day. At each station there are houses for the accommodation of passengers, with cooking-house and stables, and six or eight men always on guard. There is an established system for coolies at fixed rates, the inhabitants of the surrounding villages all taking their turn to be subject to coolie service, as well as that of guards at the station for five days at a time. This arrangement makes travelling very easy, and was a great convenience for me. I had a pleasant walk of ten or twelve miles in the morning, and the rest of the day could stroll about and explore the village and neighbourhood, having a house ready to occupy without any formalities whatever. In three days I reached Moera-dua, the first village in Rembang, and finding the country dry and undulating, with a good sprinkling of forest, I determined to remain a short time and try the neighbourhood. Just opposite the station was a small but deep river, and a good bathing-place; and beyond the village was a fine patch of forest, through which the road passed, overshadowed by magnificent trees, which partly tempted me to stay; but after a fortnight I could find no good place for insects, and very few birds different from the common species of Malacca. I therefore moved on another stage to Lobo Raman, where the guard-house is situated quite by itself in the forest, nearly a mile from each of three villages. This was very agreeable to me, as I could move about without having every motion watched by crowds of men, women and children, and I had also a much greater variety of walks to each of the villages and the plantations around them.

The villages of the Sumatran Malays are somewhat peculiar and very picturesque. A space of some acres is surrounded with a high fence, and over this area the houses are thickly strewn without the least attempt at regularity. Tall cocoa-nut trees grow abundantly between them, and the ground is bare and smooth with the trampling of many feet. The houses are raised about six feet on posts, the best being entirely built of planks, others of bamboo. The former are always more or less ornamented with carving and have high-pitched roofs and overhanging eaves. The gable ends and all the chief posts and beams are sometimes covered with exceedingly tasteful carved work, and this is still more the case in the district of Menangkabo, further west. The floor is made of split bamboo, and is rather shaky, and there is no sign of anything we should call furniture. There are no benches or chairs or stools, but merely the level floor covered with mats, on which the inmates sit or lie. The aspect of the village itself is very neat, the ground being often swept before the chief houses; but very bad odours abound, owing to there being under every house a stinking mud-hole, formed by all waste liquids and refuse matter, poured down through the floor above. In most other things Malays are tolerably clean—in some scrupulously so; and this peculiar and nasty custom, which is almost universal, arises, I have little doubt, from their having been originally a maritime and water-loving people, who built their houses on posts in the water, and only migrated gradually inland, first up the rivers and streams, and then into the dry interior. Habits which were at once so convenient and so cleanly, and which had been so long practised as to become a portion of the domestic life of the nation, were of course continued when the first settlers built their houses inland; and without a regular system of drainage, the arrangement of the villages is such that any other system would be very inconvenient.

In all these Sumatran villages I found considerable difficulty in getting anything to eat. It was not the season for vegetables, and when, after much trouble, I managed to procure some yams of a curious variety, I found them hard and scarcely eatable. Fowls were very scarce; and fruit was reduced to one of the poorest kinds of banana. The natives (during the wet season at least) live exclusively on rice, as the poorer Irish do on potatoes. A pot of rice cooked very dry and eaten with salt and red peppers, twice a day, forms their entire food during a large part of the year. This is no sign of poverty, but is simply custom; for their wives and children are loaded with silver armlets from wrist to elbow, and carry dozens of silver coins strung round their necks or suspended from their ears.

As I had moved away from Palembang, I had found the Malay spoken by the common people less and less pure, until at length it became quite unintelligible, although the continual recurrence of many well-known words assured me it was a form of Malay, and enabled me to guess at the main subject of conversation. This district had a very bad reputation a few years ago, and travellers were frequently robbed and murdered. Fights between village and village were also of frequent occurrence, and many lives were lost, owing to disputes about boundaries or intrigues with women. Now, however, since the country has been divided into districts under "Controlleurs," who visit every village in turn to hear complaints and settle disputes, such things are heard of no more. This is one of the numerous examples I have met with of the good effects of the Dutch Government. It exercises a strict surveillance over its most distant possessions, establishes a form of government well adapted to the character of the people, reforms abuses, punishes crimes, and makes itself everywhere respected by the native population.

Lobo Raman is a central point of the east end of Sumatra, being about a hundred and twenty miles from the sea to the east, north, and west. The surface is undulating, with no mountains or even hills, and there is no rock, the soil being generally a red friable clay. Numbers of small streams and rivers intersect the country, and it is pretty equally divided between open clearings and patches of forest, both virgin and second growth, with abundance of fruit trees; and there is no lack of paths to get about in any direction. Altogether it is the very country that would promise most for a naturalist, and I feel sure that at a more favourable time of year it would prove exceedingly rich; but it was now the rainy season, when, in the very best of localities, insects are always scarce, and there being no fruit on the trees, there was also a scarcity of birds. During a month's collecting, I added only three or four new species to my list of birds, although I obtained very fine specimens of many which were rare and interesting. In butterflies I was rather more successful, obtaining several fine species quite new to me, and a considerable number of very rare and beautiful insects. I will give here some account of two species of butterflies, which, though very common in collections, present us with peculiarities of the highest interest.

The first is the handsome Papilio memnon, a splendid butterfly of a deep black colour, dotted over with lines and groups of scales of a clear ashy blue. Its wings are five inches in expanse, and the hind wings are rounded, with scalloped edges. This applies to the males; but the females are very different, and vary so much that they were once supposed to form several distinct species. They may be divided into two groups—those which resemble the male in shape, and, those which differ entirely from him in the outline of the wings. The first vary much in colour, being often nearly white with dusky yellow and red markings, but such differences often occur in butterflies. The second group are much more extraordinary, and would never be supposed to be the same insect, since the hind wings are lengthened out into large spoon-shaped tails, no rudiment of which is ever to be perceived in the males or in the ordinary form of females. These tailed females are never of the dark and blue-glossed tints which prevail in the male and often occur in the females of the same form, but are invariably ornamented with stripes and patches of white or buff, occupying the larger part of the surface of the hind wings. This peculiarity of colouring led me to discover that this extraordinary female closely resembles (when flying) another butterfly of the same genus but of a different group (Papilio coön), and that we have here a case of mimicry similar to those so well illustrated and explained by Mr. Bates.[Trans. Linn. Soc. vol. xviii. p. 495; "Naturalist on the Amazons," vol. i. p. 290.]

That the resemblance is not accidental is sufficiently proved by the fact, that in the North of India, where Papilio coön is replaced by an allied form, (Papilio Doubledayi) having red spots in place of yellow, a closely-allied species or variety of Papilio memnon (P. androgeus) has the tailed female also red spotted. The use and reason of this resemblance appears to be that the butterflies imitated belong to a section of the genus Papilio which from some cause or other are not attacked by birds, and by so closely resembling these in form and colour the female of Memnon and its ally, also escape persecution. Two other species of this same section (Papilio antiphus and Papilio polyphontes) are so closely imitated by two female forms of Papilio theseus (which comes in the same section with Memnon), that they completely deceived the Dutch entomologist De Haan, and he accordingly classed them as the same species!

But the most curious fact connected with these distinct forms is that they are both the offspring of either form. A single brood of larva were bred in Java by a Dutch entomologist, and produced males as well as tailed and tailless females, and there is every reason to believe that this is always the case, and that forms intermediate in character never occur. To illustrate these phenomena, let us suppose a roaming Englishman in some remote island to have two wives—one a black-haired, red-skinned Indian, the other a woolly-headed, sooty-skinned negress; and that instead of the children being mulattoes of brown or dusky tints, mingling the characteristics of each parent in varying degrees, all the boys should be as fair-skinned and blue-eyed as their father, while the girls should altogether resemble their mothers. This would be thought strange enough, but the case of these butterflies is yet more extraordinary, for each mother is capable not only of producing male offspring like the father, and female like herself, but also other females like her fellow wife, and altogether differing from herself!

The other species to which I have to direct attention is the Kallima paralekta, a butterfly of the same family group as our Purple Emperor, and of about the same size or larger. Its upper surface is of a rich purple, variously tinged with ash colour, and across the forewings there is a broad bar of deep orange, so that when on the wing it is very conspicuous. This species was not uncommon in dry woods and thickets, and I often endeavoured to capture it without success, for after flying a short distance it would enter a bush among dry or dead leaves, and however carefully I crept up to the spot I could never discover it until it would suddenly start out again and then disappear in a similar place. If at length I was fortunate enough to see the exact spot where the butterfly settled, and though I lost sight of it for some time, I would discover that it was close before my eyes, but that in its position of repose it so closely resembled a dead leaf attached to a twig as almost certainly to deceive the eye even when gazing full upon it. I captured several specimens on the wing, and was able fully to understand the way in which this wonderful resemblance is produced.

The end of the upper wings terminates in a fine point, just as the leaves of many tropical shrubs and trees are pointed, while the lower wings are somewhat more obtuse, and are lengthened out into a short thick tail. Between these two points there runs a dark curved line exactly representing the midrib of a leaf, and from this radiate on each side a few oblique marks which well imitate the lateral veins. These marks are more clearly seen on the outer portion of the base of the wings, and on the innerside towards the middle and apex, and they are produced by striae and markings which are very common in allied species, but which are here modified and strengthened so as to imitate more exactly the venation of a leaf. The tint of the undersurface varies much, but it is always some ashy brown or reddish colour, which matches with those of dead leaves. The habit of the species is always to rest on a twig and among dead or dry leaves, and in this position with the wings closely pressed together, their outline is exactly that of a moderately-sized leaf, slightly curved or shrivelled. The tail of the hind wings forms a perfect stalk, and touches the stick while the insect is supported by the middle pair of legs, which are not noticed among the twigs and fibres that surround it. The head and antennae are drawn back between the wings so as to be quite concealed, and there is a little notch hollowed out at the very base of the wings, which allows the head to be retracted sufficiently. All these varied details combine to produce a disguise that is so complete and marvellous as to astonish everyone who observes it; and the habits of the insects are such as to utilize all these peculiarities, and render them available in such a manner as to remove all doubt of the purpose of this singular case of mimicry, which is undoubtedly a protection to the insect.

Its strong and swift flight is sufficient to save it from its enemies when on the wing, but if it were equally conspicuous when at rest it could not long escape extinction, owing to the attacks of the insectivorous birds and reptiles that abound in the tropical forests. A very closely allied species, Kallima inachis, inhabits India, where it is very common, and specimens are sent in every collection from the Himalayas. On examining a number of these, it will be seen that no two are alike, but all the variations correspond to those of dead leaves. Every tint of yellow, ash, brown, and red is found here, and in many specimens there occur patches and spots formed of small black dots, so closely resembling the way in which minute fungi grow on leaves that it is almost impossible at first not to believe that fungi have grown on the butterflies themselves!

If such an extraordinary adaptation as this stood alone, it would be very difficult to offer any explanation of it; but although it is perhaps the most perfect case of protective imitation known, there are hundreds of similar resemblances in nature, and from these it is possible to deduce a general theory of the manner in which they have been slowly brought about. The principle of variation and that of "natural selection," or survival of the fittest, as elaborated by Mr. Darwin in his celebrated "Origin of Species," offers the foundation for such a theory; and I have myself endeavoured to apply it to all the chief cases of imitation in an article published in the "Westminster Review" for 1867, entitled, "Mimicry, and other Protective Resemblances Among Animals," to which any reader is referred who wishes to know more about this subject.

In Sumatra, monkeys are very abundant, and at Lobo Kaman they used to frequent the trees which overhang the guard-house, and give me a fine opportunity of observing their gambols. Two species of Semnopithecus were most plentiful—monkeys of a slender form, with very long tails. Not being much shot at they are rather bold, and remain quite unconcerned when natives alone are present; but when I came out to look at them, they would stare for a minute or two and then make off. They take tremendous leaps from the branches of one tree to those of another a little lower, and it is very amusing when one strong leader takes a bold jump, to see the others following with more or less trepidation; and it often happens that one or two of the last seem quite unable to make up their minds to leap until the rest are disappearing, when, as if in desperation at being left alone, they throw themselves frantically into the air, and often go crashing through the slender branches and fall to the ground.

A very curious ape, the Siamang, was also rather abundant, but it is much less bold than the monkeys, keeping to the virgin forests and avoiding villages. This species is allied to the little long-armed apes of the genus Hylobates, but is considerably larger, and differs from them by having the two first fingers of the feet united together, nearly to the end as does its Latin name, Siamanga syndactyla. It moves much more slowly than the active Hylobates, keeping lower down in trees, and not indulging in such tremendous leaps; but it is still very active, and by means of its immense long arms, five feet six inches across in an adult about three feet high, can swing itself along among the trees at a great rate. I purchased a small one, which had been caught by the natives and tied up so tightly as to hurt it. It was rather savage at first, and tried to bite; but when we had released it and given it two poles under the verandah to hang upon, securing it by a short cord, running along the pole with a ring so that it could move easily, it became more contented, and would swing itself about with great rapidity. It ate almost any kind of fruit and rice, and I was in hopes to have brought it to England, but it died just before I started. It took a dislike to me at first, which I tried to get over by feeding it constantly myself. One day, however, it bit me so sharply while giving it food, that I lost patience and gave it rather a severe beating, which I regretted afterwards, as from that time it disliked me more than ever. It would allow my Malay boys to play with it, and for hours together would swing by its arms from pole to pole and on to the rafters of the verandah, with so much ease and rapidity, that it was a constant source of amusement to us. When I returned to Singapore it attracted great attention, as no one had seen a Siamang alive before, although it is not uncommon in some parts of the Malay peninsula.

As the Orangutan is known to inhabit Sumatra, and was in fact first discovered there, I made many inquiries about it; but none of the natives had ever heard of such an animal, nor could I find any of the Dutch officials who knew anything about it. We may conclude, therefore, that it does not inhabit the great forest plains in the east of Sumatra where one would naturally expect to find it, but is probably confined to a limited region in the northwest part of the island entirely in the hands of native rulers. The other great Mammalia of Sumatra, the elephant and the rhinoceros, are more widely distributed; but the former is much more scarce than it was a few years ago, and seems to retire rapidly before the spread of cultivation. Lobo Kaman tusks and bones are occasionally found about in the forest, but the living animal is now never seen. The rhinoceros (Rhinoceros sumatranus) still abounds, and I continually saw its tracks and its dung, and once disturbed one feeding, which went crashing away through the jungle, only permitting me a momentary glimpse of it through the dense underwood. I obtained a tolerably perfect cranium, and a number of teeth, which were picked up by the natives.

Another curious animal, which I had met with in Singapore and in Borneo, but which was more abundant here, is the Galeopithecus, or flying lemur. This creature has a broad membrane extending all around its body to the extremities of the toes, and to the point of the rather long tail. This enables it to pass obliquely through the air from one tree to another. It is sluggish in its motions, at least by day, going up a tree by short runs of a few feet, and then stopping a moment as if the action was difficult. It rests during the day clinging to the trunks of trees, where its olive or brown fur, mottled with irregular whitish spots and blotches, resembles closely the colour of mottled bark, and no doubt helps to protect it. Once, in a bright twilight, I saw one of these animals run up a trunk in a rather open place, and then glide obliquely through the air to another tree, on which it alighted near its base, and immediately began to ascend. I paced the distance from the one tree to the other, and found it to be seventy yards; and the amount of descent I estimated at not more than thirty-five or forty feet, or less than one in five. This I think proves that the animal must have some power of guiding itself through the air, otherwise in so long a distance it would have little chance of alighting exactly upon the trunk. Like the Cuscus of the Moluccas, the Galeopithecus feeds chiefly on leaves, and possesses a very voluminous stomach and long convoluted intestines. The brain is very small, and the animal possesses such remarkable tenacity of life, that it is exceedingly difficult to kill it by any ordinary means. The tail is prehensile; and is probably made use of as an additional support while feeding. It is said to have only a single young one at a time, and my own observation confirms this statement, for I once shot a female with a very small blind and naked little creature clinging closely to its breast, which was quite bare and much wrinkled, reminding me of the young of Marsupials, to which it seemed to form a transition. On the back, and extending over the limbs and membrane, the fur of these animals is short, but exquisitely soft, resembling in its texture that of the Chinchilla.

I returned to Palembang by water, and while staying a day at a village while a boat was being made watertight, I had the good fortune to obtain a male, female, and young bird of one of the large hornbills. I had sent my hunters to shoot, and while I was at breakfast they returned, bringing me a fine large male of the Buceros bicornis, which one of them assured me he had shot while feeding the female, which was shut up in a hole in a tree. I had often read of this curious habit, and immediately returned to the place, accompanied by several of the natives. After crossing a stream and a bog, we found a large tree leaning over some water, and on its lower side, at a height of about twenty feet, appeared a small hole, and what looked like a quantity of mud, which I was assured had been used in stopping up the large hole. After a while we heard the harsh cry of a bird inside, and could see the white extremity of its beak put out. I offered a rupee to anyone who would go up and get the bird out, with the egg or young one; but they all declared it was too difficult, and they were afraid to try. I therefore very reluctantly came away. About an hour afterwards, much to my surprise, a tremendous loud, hoarse screaming was heard, and the bird was brought me, together with a young one which had been found in the hole. This was a most curious object, as large as a pigeon, but without a particle of plumage on any part of it. It was exceedingly plump and soft, and with a semi-transparent skin, so that it looked more like a bag of jelly, with head and feet stuck on, than like a real bird.

The extraordinary habit of the male, in plastering up the female with her egg, and feeding her during the whole time of incubation, and until the young one is fledged, is common to several of the large hornbills, and is one of those strange facts in natural history which are "stranger than fiction."

CHAPTER IX. NATURAL HISTORY OF THE INDO-MALAY ISLANDS.

IN the first CHAPTER of this work I have stated generally the reasons which lead us to conclude that the large islands in the western portion of the Archipelago—Java, Sumatra, and Borneo—as well as the Malay peninsula and the Philippine islands, have been recently separated from the continent of Asia. I now propose to give a sketch of the Natural History of these, which I term the Indo-Malay islands, and to show how far it supports this view, and how much information it is able to give us of the antiquity and origin of the separate islands.

The flora of the Archipelago is at present so imperfectly known, and I have myself paid so little attention to it, that I cannot draw from it many facts of importance. The Malayan type of vegetation is however a very important one; and Dr. Hooker informs us, in his "Flora Indica," that it spreads over all the moister and more equable parts of India, and that many plants found in Ceylon, the Himalayas, the Nilghiri, and Khasia mountains are identical with those of Java and the Malay peninsula. Among the more characteristic forms of this flora are the rattans—climbing palms of the genus Calamus, and a great variety of tall, as well as stemless palms.

Orchids, Araceae, Zingiberaceae and ferns, are especially abundant, and the genus Grammatophyllum—a gigantic epiphytal orchid, whose clusters of leaves and flower-stems are ten or twelve feet long—is peculiar to it. Here, too, is the domain of the wonderful pitcher plants (Nepenthaceae), which are only represented elsewhere by solitary species in Ceylon, Madagascar, the Seychelles, Celebes, and the Moluccas. Those celebrated fruits, the Mangosteen and the Durian, are natives of this region, and will hardly grow out of the Archipelago. The mountain plants of Java have already been alluded to as showing a former connexion with the continent of Asia; and a still more extraordinary and more ancient connection with Australia has been indicated by Mr. Low's collections from the summit of Kini-balou, the loftiest mountain in Borneo.

Plants have much greater facilities for passing across arms of the sea than animals. The lighter seeds are easily carried by the winds, and many of them are specially adapted to be so carried. Others can float a long time unhurt in the water, and are drifted by winds and currents to distant shores. Pigeons, and other fruit-eating birds, are also the means of distributing plants, since the seeds readily germinate after passing through their bodies. It thus happens that plants which grow on shores and lowlands have a wide distribution, and it requires an extensive knowledge of the species of each island to determine the relations of their floras

Volume I

with any approach to accuracy. At present we have no such complete knowledge of the botany of the several islands of the Archipelago; and it is only by such striking phenomena as the occurrence of northern and even European genera on the summits of the Javanese mountains that we can prove the former connection of that island with the Asiatic continent. With land animals, however, the case is very different. Their means of passing a wide expanse of sea are far more restricted. Their distribution has been more accurately studied, and we possess a much more complete knowledge of such groups as mammals and birds in most of the islands, than we do of the plants. It is these two classes which will supply us with most of our facts as to the geographical distribution of organized beings in this region.

The number of Mammalia known to inhabit the Indo-Malay region is very considerable, exceeding 170 species. With the exception of the bats, none of these have any regular means of passing arms of the sea many miles in extent, and a consideration of their distribution must therefore greatly assist us in determining whether these islands have ever been connected with each other or with the continent since the epoch of existing species.

The Quadrumana or monkey tribe form one of the most characteristic features of this region. Twenty-four distinct species are known to inhabit it, and these are distributed with tolerable uniformity over the islands, nine being found in Java, ten in the Malay peninsula, eleven in Sumatra, and thirteen in Borneo. The great man-like Orangutans are found only in Sumatra and Borneo; the curious Siamang (next to them in size) in Sumatra and Malacca; the long-nosed monkey only in Borneo; while every island has representatives of the Gibbons or long-armed apes, and of monkeys. The lemur-like animals, Nycticebus, Tarsius, and Galeopithecus, are found on all the islands.

Seven species found on the Malay peninsula extend also into Sumatra, four into Borneo, and three into Java; while two range into Siam and Burma, and one into North India. With the exception of the Orangutan, the Siamang, the Tarsius spectrum, and the Galeopithecus, all the Malayan genera of Quadrumana are represented in India by closely allied species, although, owing to the limited range of most of these animals, so few are absolutely identical.

Of Carnivora, thirty-three species are known from the Indo-Malay region, of which about eight are found also in Burma and India. Among these are the tiger, leopard, a tiger-cat, civet, and otter; while out of the twenty genera of Malayan Carnivora, thirteen are represented in India by more or less closely allied species. As an example, the Malayan bear is represented in North India by the Tibetan bear, both of which may be seen alive at the Zoological Society's Gardens.

The hoofed animals are twenty-two in number, of which

about seven extend into Burmah and India. All the deer are of peculiar species, except two, which range from Malacca into India. Of the cattle, one Indian species reaches Malacca, while the Bos sondiacus of Java and Borneo is also found in Siam and Burma. A goat-like animal is found in Sumatra which has its representative in India; while the two-horned rhinoceros of Sumatra and the single-horned species of Java, long supposed to be peculiar to these islands, are now both ascertained to exist in Burma, Pegu, and Moulmein. The elephant of Sumatra, Borneo, and Malacca is now considered to be identical with that of Ceylon and India.

In all other groups of Mammalia the same general phenomena recur. A few species are identical with those of India. A much larger number are closely allied or representative forms, while there are always a small number of peculiar genera, consisting of animals unlike those found in any other part of the world. There are about fifty bats, of which less than one-fourth are Indian species; thirty-four Rodents (squirrels, rats, &c.), of which six or eight only are Indian; and ten Insectivora, with one exception peculiar to the Malay region. The squirrels are very abundant and characteristic, only two species out of twenty-five extending into Siam and Burma. The Tupaias are curious insect-eaters, which closely resemble squirrels, and are almost confined to the Malay islands, as are the small feather-tailed Ptilocerus lowii of Borneo, and the curious long-snouted and naked-tailed Gymnurus rafflesii.

As the Malay peninsula is a part of the continent of Asia, the question of the former union of the islands to the mainland will be best elucidated by studying the species which are found in the former district, and also in some of the islands. Now, if we entirely leave out of consideration the bats, which have the power of flight, there are still forty-eight species of mammals common to the Malay peninsula and the three large islands. Among these are seven Quadrumana (apes, monkeys, and lemurs), animals who pass their whole existence in forests, who never swim, and who would be quite unable to traverse a single mile of sea; nineteen Carnivora, some of which no doubt might cross by swimming, but we cannot suppose so large a number to have passed in this way across a strait which, except at one point, is from thirty to fifty miles wide; and five hoofed animals, including the Tapir, two species of rhinoceros, and an elephant. Besides these there are thirteen Rodents and four Insectivora, including a shrew-mouse and six squirrels, whose unaided passage over twenty miles of sea is even more inconceivable than that of the larger animals.

But when we come to the cases of the same species inhabiting two of the more widely separated islands, the difficulty is much increased. Borneo is distant nearly 150 miles from Biliton, which is about fifty miles from Banca, and this fifteen from Sumatra, yet there are no less than thirty-six species of mammals common to Borneo and

Sumatra. Java again is more than 250 miles from Borneo, yet these two islands have twenty-two species in common, including monkeys, lemurs, wild oxen, squirrels and shrews. These facts seem to render it absolutely certain that there has been at some former period a connection between all these islands and the mainland, and the fact that most of the animals common to two or more of then, show little or no variation, but are often absolutely identical, indicates that the separation must have been recent in a geological sense; that is, not earlier than the Newer Pliocene epoch, at which time land animals began to assimilate closely with those now existing.

Even the bats furnish an additional argument, if one were needed, to show that the islands could not have been peopled from each other and from the continent without some former connection. For if such had been the mode of stocking them with animals, it is quite certain that creatures which can fly long distances would be the first to spread from island to island, and thus produce an almost perfect uniformity of species over the whole region. But no such uniformity exists, and the bats of each island are almost, if not quite, as distinct as the other mammals. For example, sixteen species are known in Borneo, and of these ten are found in Java and five in Sumatra, a proportion about the same as that of the Rodents, which have no direct means of migration. We learn from this fact, that the seas which separate the islands from each other are wide enough to prevent the passage even of flying animals, and that we must look to the same causes as having led to the present distribution of both groups. The only sufficient cause we can imagine is the former connection of all the islands with the continent, and such a change is in perfect harmony with what we know of the earth's past history, and is rendered probable by the remarkable fact that a rise of only three hundred feet would convert the wide seas that separate them into an immense winding valley or plain about three hundred miles wide and twelve hundred long. It may, perhaps, be thought that birds which possess the power of flight in so pre-eminent a degree, would not be limited in their range by arms of the sea, and would thus afford few indications of the former union or separation of the islands they inhabit. This, however, is not the case. A very large number of birds appear to be as strictly limited by watery barriers as are quadrupeds; and as they have been so much more attentively collected, we have more complete materials to work upon, and are able to deduce from them still more definite and satisfactory results. Some groups, however, such as the aquatic birds, the waders, and the birds of prey, are great wanderers; other groups are little known except to ornithologists. I shall therefore refer chiefly to a few of the best known and most remarkable families of birds as a sample of the conclusions furnished by the entire class.

Volume I

The birds of the Indo-Malay region have a close resemblance to those of India; for though a very large proportion of the species are quite distinct, there are only about fifteen peculiar genera, and not a single family group confined to the former district. If, however, we compare the islands with the Burmese, Siamese, and Malayan countries, we shall find still less difference, and shall be convinced that all are closely united by the bond of a former union. In such well-known families as the woodpeckers, parrots, trogons, barbets, kingfishers, pigeons, and pheasants, we find some identical species spreading over all India, and as far as Java and Borneo, while a very large proportion are common to Sumatra and the Malay peninsula.

The force of these facts can only be appreciated when we come to treat the islands of the Austro-Malay region, and show how similar barriers have entirely prevented the passage of birds from one island to another, so that out of at least three hundred and fifty land birds inhabiting Java and Borneo, not more than ten have passed eastward into Celebes. Yet the Straits of Macassar are not nearly so wide as the Java sea, and at least a hundred species are common to Borneo and Java.

I will now give two examples to show how a knowledge of the distribution of animals may reveal unsuspected facts in the past history of the earth. At the eastern extremity of Sumatra, and separated from it by a strait about fifteen miles wide, is the small rocky island of Banca, celebrated for its tin mines. One of the Dutch residents there sent some collections of birds and animals to Leyden, and among them were found several species distinct from those of the adjacent coast of Sumatra. One of these was a squirrel (Sciurus bangkanus), closely allied to three other species inhabiting respectively the Malay peninsula, Sumatra, and Borneo, but quite as distinct from them all as they are from each other. There were also two new ground thrushes of the genus Pitta, closely allied to, but quite distinct from, two other species inhabiting both Sumatra and Borneo, and which did not perceptibly differ in these large and widely separated islands. This is just as if the Isle of Man possessed a peculiar species of thrush and blackbird, distinct from the birds which are common to England and Ireland.

These curious facts would indicate that Banca may have existed as a distinct island even longer than Sumatra and Borneo, and there are some geological and geographical facts which render this not so improbable as it would at first seem to be. Although on the map Banca appears so close to Sumatra, this does not arise from its having been recently separated from it; for the adjacent district of Palembang is new land, being a great alluvial swamp formed by torrents from the mountains a hundred miles distant.

Banca, on the other hand, agrees with Malacca, Singapore, and the intervening island of Lingen, in being formed of granite and laterite; and these have all most likely once formed an extension of the Malay peninsula. As the rivers of Borneo and Sumatra have been for ages filling up the intervening sea, we may be sure that its depth has recently been greater, and it is very probable that those large islands were never directly connected with each other except through the Malay peninsula. At that period the same species of squirrel and Pitta may have inhabited all these countries; but when the subterranean disturbances occurred which led to the elevation of the volcanoes of Sumatra, the small island of Banca may have been separated first, and its productions being thus isolated might be gradually modified before the separation of the larger islands had been completed.

As the southern part of Sumatra extended eastward and formed the narrow straits of Banca, many birds and insects and some Mammalia would cross from one to the other, and thus produce a general similarity of productions, while a few of the older inhabitants remained, to reveal by their distinct forms, their different origin. Unless we suppose some such changes in physical geography to have occurred, the presence of peculiar species of birds and mammals in such an island as Banca is a hopeless puzzle; and I think I have shown that the changes required are by no means so improbable as a mere glance at the map would lead us to suppose.

For our next example let us take the great islands of Sumatra and Java. These approach so closely together, and the chain of volcanoes that runs through them gives such an air of unity to the two, that the idea of their having been recently dissevered is immediately suggested. The natives of Java, however, go further than this; for they actually have a tradition of the catastrophe which broke them asunder, and fix its date at not much more than a thousand years ago. It becomes interesting, therefore, to see what support is given to this view by the comparison of their animal productions.

The Mammalia have not been collected with sufficient completeness in both islands to make a general comparison of much value, and so many species have been obtained only as live specimens in captivity, that their locality has often been erroneously given, the island in which they were obtained being substituted for that from which they originally came. Taking into consideration only those whose distribution is more accurately known, we learn that Sumatra is, in a zoological sense, more nearly related to Borneo than it is to Java. The great man-like apes, the elephant, the tapir, and the Malay bear, are all common to the two former countries, while they are absent from the latter. Of the three long-tailed monkeys (Semnopithecus) inhabiting Sumatra, one extends into Borneo, but the two species of Java are both peculiar to it. So also the great Malay deer (Rusa equina), and the small Tragulus kanchil, are common to Sumatra and Borneo, but do not extend into Java, where they are replaced by Tragulas javanicus. The tiger, it is true, is found in Sumatra and Java, but not in Borneo. But as this animal is known to swim well, it may have found its way across the Straits of Sunda, or it may have inhabited Java before it was separated from the mainland, and from some unknown cause have ceased to exist in Borneo.

In Ornithology there is a little uncertainty owing to the birds of Java and Sumatra being much better known than those of Borneo; but the ancient separation of Java as an island is well exhibited by the large number of its species which are not found in any of the other islands. It possesses no less than seven pigeons peculiar to itself, while Sumatra has only one. Of its two parrots one extends into Borneo, but neither into Sumatra. Of the fifteen species of woodpeckers inhabiting Sumatra only four reach Java, while eight of them are found in Borneo and twelve in the Malay peninsula. The two Trogons found in Java are peculiar to it, while of those inhabiting Sumatra at least two extend to Malacca and one to Borneo. There are a very large number of birds, such as the great Argus pheasant, the fire-backed and ocellated pheasants, the crested partridge (Rollulus coronatus), the small Malacca parrot (Psittinus incertus), the great helmeted hornbill (Buceroturus galeatus), the pheasant ground-cuckoo (Carpococcyx radiatus), the rose-crested bee-eater (Nyctiornis amicta), the great gaper (Corydon sumatranus), and the green-crested gaper (Calyptomena viridis), and many others, which are common to Malacca, Sumatra, and Borneo, but are entirely absent from Java. On the other hand we have the peacock, the green jungle cock, two blue ground thrushes (Arrenga cyanea and Myophonus flavirostris), the fine pink-headed dove (Ptilonopus porphyreus), three broad-tailed ground pigeons (Macropygia), and many other interesting birds, which are found nowhere in the Archipelago out of Java.

Insects furnish us with similar facts wherever sufficient data are to be had, but owing to the abundant collections that have been made in Java, an unfair preponderance may be given to that island. This does not, however, seem to be the case with the true Papilionidae or swallow-tailed butterflies, whose large size and gorgeous colouring has led to their being collected more frequently than other insects. Twenty-seven species are known from Java, twenty-nine from Borneo, and only twenty-one from Sumatra. Four are entirely confined to Java, while only two are peculiar to Borneo and one to Sumatra. The isolation of Java will, however, be best shown by grouping the islands in pairs, and indicating the number of species common to each pair. Thus:—

Borneo .. . .. 29 species
Sumatra.. . .. 21 do.   20 species common to both islands.

Borneo .. . .. 29 do.
Java. .. . .. 27 do.   20   do.   do.

Sumatra.. . .. 21 do.
Java. .. . .. 27 do.   11   do.   do.

Making some allowance for our imperfect knowledge of the Sumatran species, we see that Java is more isolated from the two larger islands than they are from each other, thus entirely confirming the results given by the distribution of birds and Mammalia, and rendering it almost certain that the last-named island was the first to be completely separated from the Asiatic continent, and that the native tradition of its having been recently separated from Sumatra is entirely without foundation.

We are now able to trace out with some probability the course of events. Beginning at the time when the whole of the Java sea, the Gulf of Siam, and the Straits of Malacca were dry land, forming with Borneo, Sumatra, and Java, a vast southern prolongation of the Asiatic continent, the first movement would be the sinking down of the Java sea, and the Straits of Sunda, consequent on the activity of the Javanese volcanoes along the southern extremity of the land, and leading to the complete separation of that island. As the volcanic belt of Java and Sumatra increased in activity, more and more of the land was submerged, until first Borneo, and afterwards Sumatra, became entirely severed. Since the epoch of the first disturbance, several distinct elevations and depressions may have taken place, and the islands may have been more than once joined with each other or with the main land, and again separated. Successive waves of immigration may thus have modified their animal productions, and led to those anomalies in distribution which are so difficult to account for by any single operation of elevation or submergence. The form of Borneo, consisting of radiating mountain chains with intervening broad alluvial valleys, suggests the idea that it has once been much more submerged than it is at present (when it would have somewhat resembled Celebes or Gilolo in outline), and has been increased to its present dimensions by the filling up of its gulfs with sedimentary matter, assisted by gradual elevation of the land. Sumatra has also been evidently much increased in size by the formation of alluvial plains along its northeastern coasts.

There is one peculiarity in the productions of Java that is very puzzling—the occurrence of several species or groups characteristic of the Siamese countries or of India, but which do not occur in Borneo or Sumatra. Among

Volume I

Mammals the Rhinoceros javanicus is the most striking example, for a distinct species is found in Borneo and Sumatra, while the Javanese species occurs in Burma and even in Bengal. Among birds, the small ground-dove, Geopelia striata, and the curious bronze-coloured magpie, Crypsirhina varians, are common to Java and Siam; while there are in Java species of Pteruthius, Arrenga, Myiophonus, Zoothera, Sturnopastor, and Estrelda, the near allies of which are found in various parts of India, while nothing like them is known to inhabit Borneo or Sumatra.

Such a curious phenomenon as this can only be understood by supposing that, subsequent to the separation of Java, Borneo became almost entirely submerged, and on its re-elevation was for a time connected with the Malay peninsula and Sumatra, but not with Java or Siam. Any geologist who knows how strata have been contorted and tilted up, and how elevations and depressions must often have occurred alternately, not once or twice only, but scores and even hundreds of times, will have no difficulty in admitting that such changes as have been here indicated, are not in themselves improbable. The existence of extensive coal-beds in Borneo and Sumatra, of such recent origin that the leaves which abound in their shales are scarcely distinguishable from those of the forests which now cover the country, proves that such changes of level actually did take place; and it is a matter of much interest, both to the geologist and to the philosophic naturalist, to be able to form some conception of the order of those changes, and to understand how they may have resulted in the actual distribution of animal life in these countries; a distribution which often presents phenomena so strange and contradictory, that without taking such changes into consideration we are unable even to imagine how they could have been brought about.

CHAPTER X. BALI AND LOMBOCK.

(JUNE, JULY, 1856.)

THE islands of Bali and Lombock, situated at the eastern end of Java, are particularly interesting. They are the only islands of the whole Archipelago in which the Hindu religion still maintains itself—and they form the extreme points of the two great zoological divisions of the Eastern hemisphere; for although so similar in external appearance and in all physical features, they differ greatly in their natural productions. It was after having spent two years in Borneo, Malacca and Singapore, that I made a somewhat involuntary visit to these islands on my way to Macassar. Had I been able to obtain a passage direct to that place from Singapore, I should probably never have gone near them, and should have missed some of the most important discoveries of my whole expedition the East.

It was on the 13th of June, 1856, after a twenty days' passage from Singapore in the "Kembang Djepoon" (Rose of Japan), a schooner belonging to a Chinese merchant, manned by a Javanese crew, and commanded by an English captain, that we cast anchor in the dangerous roadstead of Bileling on the north side of the island of Bali. Going on shore with the captain and the Chinese supercargo, I was at once introduced to a novel and interesting scene. We went first to the house of the Chinese Bandar, or chief merchant, where we found a number of natives, well dressed, and all conspicuously armed with krisses, displaying their large handles of ivory or gold, or beautifully grained and polished wood.

The Chinamen had given up their national costume and adopted the Malay dress, and could then hardly be distinguished from the natives of the island—an indication of the close affinity of the Malayan and Mongolian races. Under the thick shade of some mango-trees close by the house, several women-merchants were selling cotton goods; for here the women trade and work for the benefit of their husbands, a custom which Mahometan Malays never adopt. Fruit, tea, cakes, and sweetmeats were brought to us; many questions were asked about our business and the state of trade in Singapore, and we then took a walk to look at the village. It was a very dull and dreary place; a collection of narrow lanes bounded by high mud walls, enclosing bamboo houses, into some of which we entered and were very kindly received.

During the two days that we remained here, I walked out into the surrounding country to catch insects, shoot birds, and spy out the nakedness or fertility of the land. I was both astonished and delighted; for as my visit to Java was some years later, I had never beheld so beautiful and well cultivated a district out of Europe. A slightly undulating plain extends from the seacoast about ten or twelve miles inland, where it is bounded by a wide range of wooded and cultivated hills. Houses and villages, marked out by dense clumps of cocoa-nut palms, tamarind and other fruit trees, are dotted about in every direction; while between them extend luxuriant rice-grounds, watered by an elaborate system of irrigation that would be the pride of the best cultivated parts of Europe. The whole surface of the country is divided into irregular patches, following the undulations of the ground, from many acres to a few perches in extent, each of which is itself perfectly level, but stands a few inches or several feet above or below those adjacent to it. Every one of these patches can be flooded or drained at will by means of a system of ditches and small channels, into which are diverted the whole of the streams that descend from the mountains. Every patch now bore crops in various stages of growth, some almost ready for cutting, and all in the most flourishing condition and of the most exquisite green tints.

The sides of the lanes and bridle roads were often edged with prickly Cacti and a leafless Euphorbia, but the country being so highly cultivated there was not much room for indigenous vegetation, except upon the sea-beach. We saw plenty of the fine race of domestic cattle descended from the Bos banteng of Java, driven by half naked boys, or tethered in pasture-grounds. They are large and handsome animals, of a light brown colour, with white legs, and a conspicuous oval patch behind of the same colour. Wild cattle of the same race are said to be still found in the mountains. In so well-cultivated a country it was not to be expected that I could do much in natural history, and my ignorance of how important a locality this was for the elucidation of the geographical distribution of animals, caused me to neglect obtaining some specimens which I never met with again. One of these was a weaver bird with a bright yellow head, which built its bottle-shaped nests by dozens on some trees near the beach. It was the Ploceus hypoxantha, a native of Java; and here, at the extreme limits of its range westerly, I shot and preserved specimens of a wagtail-thrush, an oriole, and some starlings, all species found in Java, and some of them peculiar to that island. I also obtained some beautiful butterflies, richly marked with black and orange on a white ground, and which were the most abundant insects in the country lanes. Among these was a new species, which I have named Pieris tamar.

Leaving Bileling, a pleasant sail of two days brought us to Ampanam in the island of Lombock, where I proposed to remain till I could obtain a passage to Macassar. We enjoyed superb views of the twin volcanoes of Bali and Lombock, each about eight thousand feet high, which form magnificent objects at sunrise and sunset, when they rise out of the mists and clouds that surround their bases, glowing with the rich and changing tints of these the most charming moments in a tropical day.

The bay or roadstead of Ampanam is extensive, and being at this season sheltered from the prevalent southeasterly winds, was as smooth as a lake. The beach of black volcanic sand is very steep, and there is at all times, a heavy surf upon it, which during spring-tides increases to such an extent that it is often impossible for boats to land, and many serious accidents have occurred. Where we lay anchored, about a quarter of a mile from the shore, not the slightest swell was perceptible, but on approaching nearer undulations began, which rapidly increased, so as to form rollers which toppled over onto the beach at regular intervals with a noise like thunder. Sometimes this surf increases suddenly during perfect calms to as great a force and fury as when a gale of wind is blowing, beating to pieces all boats that may not have been hauled sufficiently high upon the beach, and carrying away uncautious natives. This violent surf is probably in some way dependent upon the swell of the great southern ocean and the violent currents that flow through the Straits of Lombock. These are so uncertain that vessels preparing to anchor in the bay are sometimes suddenly swept away into the straits, and are not able to get back again for a fortnight.

What seamen call the "ripples" are also very violent in

the straits, the sea appearing to boil and foam and dance like the rapids below a cataract; vessels are swept about helplessly, and small ones are occasionally swamped in the finest weather and under the brightest skies.

I felt considerably relieved when all my boxes and myself had passed in safety through the devouring surf, which the natives look upon with some pride, saying, that "their sea is always hungry, and eats up everything it can catch." I was kindly received by Mr. Carter, an Englishman, who is one of the Bandars or licensed traders of the port, who offered me hospitality and every assistance during my stay. His house, storehouses, and offices were in a yard surrounded by a tall bamboo fence, and were entirely constructed of bamboo with a thatch of grass, the only available building materials. Even these were now very scarce, owing to the great consumption in rebuilding the place since the great fire some months before, which in an hour or two had destroyed every building in the town.

The next day I went to see Mr. S., another merchant to whom I had brought letters of introduction, and who lived about seven miles off. Mr. Carter kindly lent me a horse, and I was accompanied by a young Dutch gentleman residing at Ampanam, who offered to be my guide. We first passed through the town and suburbs along a straight road bordered by mud walls and a fine avenue of lofty trees; then through rice-fields, irrigated in the same manner as I had seen them at Bileling; and afterwards over sandy pastures near the sea, and occasionally along the beach itself. Mr. S. received us kindly, and offered me a residence at his house should I think the neighbourhood favourable for my pursuits. After an early breakfast we went out to explore, taking guns and insect nets. We reached some low hills which seemed to offer the most favourable ground, passing over swamps, sandy flats overgrown with coarse sedges, and through pastures and cultivated grounds, finding however very little in the way of either birds or insects. On our way we passed one or two human skeletons, enclosed within a small bamboo fence, with the clothes, pillow, mat, and betel-box of the unfortunate individual, who had been either murdered or executed. Returning to the house, we found a Balinese chief and his followers on a visit. Those of higher rank sat on chairs, the others squatted on the floor. The chief very coolly asked for beer and brandy, and helped himself and his followers, apparently more out of curiosity than anything else as regards the beer, for it seemed very distasteful to them, while they drank the brandy in tumblers with much relish.

Returning to Ampanam, I devoted myself for some days to shooting the birds of the neighbourhood. The fine fig-trees of the avenues, where a market was held, were tenanted by superb orioles (Oriolus broderpii) of a rich orange colour, and peculiar to this island and the adjacent ones of Sumbawa and Flores. All round the town were abundance of the curious Tropidorhynchus timoriensis, allied to the Friar bird of Australia. They are here called "Quaich-quaich," from their strange loud voice, which seems to repeat these words in various and not unmelodious intonations.

Every day boys were to be seen walking along the roads and by the hedges and ditches, catching dragonflies with birdlime. They carry a slender stick, with a few twigs at the end well annointed, so that the least touch captures the insect, whose wings are pulled off before it is consigned to a small basket. The dragon-flies are so abundant at the time of the rice flowering that thousands are soon caught in this way. The bodies are fried in oil with onions and preserved shrimps, or sometimes alone, and are considered a great delicacy. In Borneo, Celebes, and many other islands, the larvae of bees and wasps are eaten, either alive as pulled out of the cells, or fried like the dragonflies. In the Moluccas the grubs of the palm-beetles (Calandra) are regularly brought to market in bamboos and sold for food; and many of the great horned Lamellicorn beetles are slightly roasted on the embers and eaten whenever met with. The superabundance of insect life is therefore turned to some account by these islanders.

Finding that birds were not very numerous, and hearing much of Labuan Tring at the southern extremity of the bay, where there was said to be much uncultivated country and plenty of birds as well as deer and wild pigs, I determined to go there with my two servants, Ali, the Malay lad from Borneo, and Manuel, a Portuguese of Malacca accustomed to bird-skinning. I hired a native boat with outriggers to take us with our small quantity of luggage, and a day's rowing and tracking along the shore brought us to the place.

I had a note of introduction to an Amboynese Malay, and obtained the use of part of his house to live and work in. His name was "Inchi Daud" (Mr. David), and he was very civil; but his accommodations were limited, and he could only hire me part of his reception-room. This was the front part of a bamboo house (reached by a ladder of about six rounds very wide apart), and having a beautiful view over the bay. However, I soon made what arrangements were possible, and then set to work. The country around was pretty and novel to me, consisting of abrupt volcanic hills enclosing flat valleys or open plains. The hills were covered with a dense scrubby bush of bamboos and prickly trees and shrubs, the plains were adorned with hundreds of noble palm-trees, and in many places with a luxuriant shrubby vegetation. Birds were plentiful and very interesting, and I now saw for the first time many Australian forms that are quite absent from the islands westward. Small white cockatoos were abundant, and their loud screams, conspicuous white colour, and pretty yellow crests, rendered them a very important feature in the landscape. This is the most westerly point on the globe where any of the family are to be found. Some small honeysuckers of the genus Ptilotis, and the strange moundmaker (Megapodius gouldii), are also here first met with on the traveller's journey eastward. The last mentioned bird requires a fuller notice.

The Megapodidae are a small family of birds found only in Australia and the surrounding islands, but extending as far as the Philippines and Northwest Borneo. They are allied to the gallinaceous birds, but differ from these and from all others in never sitting upon their eggs, which they bury in sand, earth, or rubbish, and leave to be hatched by the heat of the sun or by fermentation. They are all characterised by very large feet and long curved claws, and most of the species of Megapodius rake and scratch together all kinds of rubbish, dead leaves, sticks, stones, earth, rotten wood, etc., until they form a large mound, often six feet high and twelve feet across, in the middle of which they bury their eggs. The natives can tell by the condition of these mounds whether they contain eggs or not; and they rob them whenever they can, as the brick-red eggs (as large as those of a swan) are considered a great delicacy. A number of birds are said to join in making these mounds and lay their eggs together, so that sometimes forty or fifty may be found. The mounds are to be met with here and there in dense thickets, and are great puzzles to strangers, who cannot understand who can possibly have heaped together cartloads of rubbish in such out-of-the-way places; and when they inquire of the natives they are but little wiser, for it almost always appears to them the wildest romance to be told that it is all done by birds. The species found in Lombock is about the size of a small hen, and entirely of dark olive and brown tints. It is a miscellaneous feeder, devouring fallen fruits, earthworms, snails, and centipedes, but the flesh is white and well-flavoured when properly cooked.

The large green pigeons were still better eating, and were much more plentiful. These fine birds, exceeding our largest tame pigeons in size, abounded on the palm-trees, which now bore huge bunches of fruits—mere hard globular nuts, about an inch in diameter, and covered with a dry green skin and a very small portion of pulp. Looking at the pigeon's bill and head, it would seem impossible that it could swallow such large masses, or that it could obtain any nourishment from them; yet I often shot these birds with several palm-fruits in the crop, which generally burst when they fell to the ground. I obtained here eight species of Kingfishers; among which was a very beautiful new one, named by Mr. Gould, Halcyon fulgidus. It was found always in thickets, away from water, and seemed to feed on snails and insects picked up from the ground after the manner of the great Laughing Jackass of Australia. The beautiful little violet and orange species (Ceyx rufidorsa) is found in similar situations, and darts rapidly along like a flame of fire. Here also I first met with the pretty Australian Bee-eater (Merops ornatus). This elegant little bird sits on twigs in open places, gazing eagerly around, and darting off at intervals to seize some insect which it sees flying near; returning afterwards to the same twig to swallow it. Its long, sharp, curved bill, the two long narrow feathers in its tail, its beautiful green plumage varied with rich brown and black and vivid blue on the throat, render it one of the most graceful and interesting objects a naturalist can see for the first time.

Of all the birds of Lombock, however, I sought most after the beautiful ground thrushes (Pitta concinna), and always thought myself lucky if I obtained one. They were found only in the dry plains densely covered with thickets, and carpeted at this season with dead leaves. They were so shy that it was very difficult to get a shot at them, and it was only after a good deal of practice that I discovered how to do it. The habit of these birds is to hop about on the ground, picking up insects, and on the least alarm to run into the densest thicket or take a flight close to the ground. At intervals they utter a peculiar cry of two notes which when once heard is easily recognised, and they can also be heard hopping along among the dry leaves.

My practice was, therefore, to walk cautiously along the narrow pathways with which the country abounded, and on detecting any sign of a Pitta's vicinity to stand motionless and give a gentle whistle occasionally, imitating the notes as near as possible. After half an hour's waiting I was often rewarded by seeing the pretty bird hopping along in the thicket. Then I would perhaps lose sight of it again, until having my gun raised and ready for a shot, a second glimpse would enable me to secure my prize, and admire its soft puffy plumage and lovely colours. The upper part is rich soft green, the head jet black with a stripe of blue and brown over each eye; at the base of the tail and on the shoulders are bands of bright silvery blue; the under side is delicate buff with a stripe of rich crimson, bordered with black on the belly. Beautiful grass-green doves, little crimson and black flower-peckers, large black cuckoos, metallic king-crows, golden orioles, and the fine jungle-cocks—the origin of all our domestic breeds of poultry—were among the birds that chiefly attracted my attention during our stay at Labuan Tring.

The most characteristic feature of the jungle was its thorniness. The shrubs were thorny; the creepers were thorny; the bamboos even were thorny. Everything grew zigzag and jagged, and in an inextricable tangle, so that to get through the bush with gun or net or even spectacles, was generally not to be done, and insect-catching in such localities was out of the question. It was in such places that the Pittas often lurked, and when shot it became a matter of some difficulty to secure the bird, and seldom without a heavy payment of pricks and scratches and torn clothes could the prize be won. The dry volcanic soil and arid climate seem favourable to the production of such stunted and thorny vegetation, for the natives assured me that this was nothing to the thorns and prickles of Sumbawa whose surface still bears the covering of volcanic ashes thrown out forty years ago by the terrible eruption of Tomboro.

Among the shrubs and trees that are not prickly the Apocynaceae were most abundant, their bilobed fruits of varied form and colour and often of most tempting appearance, hanging everywhere by the waysides as if to invite to destruction the weary traveller who may be unaware of their poisonous properties. One in particular with a smooth shining skin of a golden orange colour rivals in appearance the golden apples of the Hesperides, and has great attractions for many birds, from the white cockatoos to the little yellow Zosterops, who feast on the crimson seeds which are displayed when the fruit bursts open. The great palm called "Gubbong" by the natives, a species of Corypha, is the most striking feature of the plains, where it grows by thousands and appears in three different states—in leaf, in flower and fruit, or dead. It has a lofty cylindrical stem about a hundred feet high and two to three feet in diameter; the leaves are large and fan-shaped, and fall off when the tree flowers, which it does only once in its life in a huge terminal spike, upon which are produced masses of a smooth round fruit of a green colour and about an inch in diameter. When these ripen and fall the tree dies, and remains standing a year or two before it falls. Trees in leaf only are by far the most numerous, then those in flower and fruit, while dead trees are scattered here and there among them. The trees in fruit are the resort of the great green fruit pigeons, which have been already mentioned. Troops of monkeys (Macacus cynomolgus) may often be seen occupying a tree, showering down the fruit in great profusion, chattering when disturbed and making an enormous rustling as they scamper off among the dead palm leaves; while the pigeons have a loud booming voice more like the roar of a wild beast than the note of a bird.

My collecting operations here were carried on under more than usual difficulties. One small room had to serve for eating, sleeping and working, and one for storehouse and dissecting-room; in it were no shelves, cupboards, chairs or tables; ants swarmed in every part of it, and dogs, cats and fowls entered it at pleasure. Besides this it was the parlour and reception-room of my host, and I was obliged to consult his convenience and that of the numerous guests who visited us. My principal piece of furniture was a box, which served me as a dining table, a seat while skinning birds, and as the receptacle of the birds when skinned and dried. To keep them free from ants we borrowed, with some difficulty, an old bench, the four legs of which being placed in cocoa-nut shells filled with water kept us tolerably free from these pests. The box and the bench were, however, literally the only places where anything could be put away, and they were generally well occupied by two insect boxes and about a hundred birds' skins in process of drying. It may therefore be easily conceived that when anything bulky or out of the common way was collected, the question "Where is it to be put?" was rather a difficult one to answer. All animal substances moreover require some time to dry thoroughly, emit a very disagreeable odour while doing so, and are particularly attractive to ants, flies, dogs, rats, cats, and other vermin, calling for special cautions and constant supervision, which under the circumstances above described were impossible.

My readers may now partially understand why a travelling naturalist of limited means, like myself, does so much less than is expected or than he would himself wish to do. It would be interesting to preserve skeletons of many birds and animals, reptiles and fishes in spirits, skins of the larger animals, remarkable fruits and woods and the most curious articles of manufacture and commerce; but it will be seen that under the circumstances I have just described, it would have been impossible to add these to the collections which were my own more especial favourites. When travelling by boat the difficulties are as great or greater, and they are not diminished when the journey is by land. It was absolutely necessary therefore to limit my collections to certain groups to which I could devote constant personal attention, and thus secure from destruction or decay what had been often obtained by much labour and pains.

While Manuel sat skinning his birds of an afternoon, generally surrounded by a little crowd of Malays and Sassaks (as the indigenes of Lombock are termed), he often held forth to them with the air of a teacher, and was listened to with profound attention. He was very fond of discoursing on the "special providences" of which he believed he was daily the subject. "Allah has been merciful today," he would say—for although a Christian he adopted the Mahometan mode of speech—"and has given us some very fine birds; we can do nothing without him." Then one of the Malays would reply, "To be sure, birds are like mankind; they have their appointed time to die; when that time comes nothing can save them, and if it has not come you cannot kill them." A murmur of assent follow, until sentiments and cries of "Butul! Butul!" (Right, right.) Then Manuel would tell a long story of one of his unsuccessful hunts—how he saw some fine bird and followed it a long way, and then missed it, and again found it, and shot two or three times at it, but could never hit it, "Ah!" says an old Malay, "its time was not come, and so it was impossible for you to kill it." A doctrine is this which is very consoling to the bad marksman, and which quite accounts for the facts, but which is yet somehow not altogether satisfactory.

It is universally believed in Lombock that some men have the power to turn themselves into crocodiles, which they do for the sake of devouring their enemies, and many strange tales are told of such transformations. I was therefore rather surprised one evening to hear the following curious fact stated, and as it was not contradicted by any of the persons present, I am inclined to accept it provisionally as a contribution to the Natural History of the island. A Bornean Malay who had been for many years resident here said to Manuel, "One thing is strange in this country—the scarcity of ghosts." "How so?" asked Manuel. "Why, you know," said the Malay, "that in our countries to the westward, if a man dies or is killed, we dare not pass near the place at night, for all sorts of noises are heard which show that ghosts are about. But here there are numbers of men killed, and their bodies lie unburied in the fields and by the roadside, and yet you can walk by them at night and never hear or see anything at all, which is not the case in our country, as you know very well." "Certainly I do," said Manuel; and so it was settled that ghosts were very scarce, if not altogether unknown in Lombock. I would observe, however, that as the evidence is purely negative we should be wanting in scientific caution if we accepted this fact as sufficiently well established.

One evening I heard Manuel, Ali, and a Malay man whispering earnestly together outside the door, and could distinguish various allusions to "krisses," throat-cutting, heads, etc. etc. At length Manuel came in, looking very solemn and frightened, and said to me in English, "Sir—

must take care,—no safe here;—want cut throat." On further inquiry, I found that the Malay had been telling them that the Rajah had just sent down an order to the village, that they were to get a certain number of heads for an offering in the temples to secure a good crop of rice. Two or three other Malays and Bugis, as well as the Amboyna man in whose house we lived, confirmed this account, and declared that it was a regular thing every year, and that it was necessary to keep a good watch and never go out alone. I laughed at the whole thing, and tried to persuade them that it was a mere tale, but to no effect. They were all firmly persuaded that their lives were in danger. Manuel would not go out shooting alone, and I was obliged to accompany him every morning, but I soon gave him the slip in the jungle. Ali was afraid to go and look for firewood without a companion, and would not even fetch water from the well a few yards behind the house unless armed with an enormous spear. I was quite sure all the time that no such order had been sent or received, and that we were in perfect safety. This was well shown shortly afterwards, when an American sailor ran away from his ship on the east side of the island, and made his way on foot and unarmed across to Ampanam, having met with the greatest hospitality on the whole route. Nowhere would the smallest payment be taken for the food and lodging which were willingly furbished him. On pointing out this fact to Manuel, he replied, "He one bad man,—run away from his ship—no one can believe word he say;" and so I was obliged to leave him in the uncomfortable persuasion that he might any day have his throat cut.

A circumstance occurred here which appeared to throw some light on the cause of the tremendous surf at Ampanam. One evening I heard a strange rumbling noise, and at the same time the house shook slightly. Thinking it might be thunder, I asked, "What is that?" "It is an earthquake," answered Inchi Daud, my host; and he then told me that slight shocks were occasionally felt there, but he had never known them to be severe. This happened on the day of the last quarter of the moon, and consequently when tides were low and the surf usually at its weakest. On inquiry afterwards at Ampanam, I found that no earthquake had been noticed, but that on one night there had been a very heavy surf, which shook the house, and the next day there was a very high tide, the water having flooded Mr. Carter's premises, higher than he had ever known it before. These unusual tides occur every now and then, and are not thought much of; but by careful inquiry I ascertained that the surf had occurred on the very night I had felt the earthquake at Labuan Tring, nearly twenty miles off. This would seem to indicate, that although the ordinary heavy surf may be due to the swell of the great Southern Ocean confined in a narrow channel, combined with a peculiar form of bottom near the shore, yet the sudden heavy surfs and high tides that occur occasionally in perfectly calm weather, may be due to slight upheavals of the ocean-bed in this eminently volcanic region.

CHAPTER XI. LOMBOCK: MANNERS AND CUSTOMS OF THE PEOPLE.

HAVING made a very fine and interesting collection of the birds of Labuan Tring, I took leave of my kind host, Inchi Daud, and returned to Ampanam to await an opportunity to reach Macassar. As no vessel had arrived bound for that port, I determined to make an excursion into the interior of the island, accompanied by Mr. Ross, an Englishman born in the Keeling Islands, and now employed by the Dutch Government to settle the affairs of a missionary who had unfortunately become bankrupt here. Mr. Carter kindly lent me a horse, and Mr. Ross took his native groom.

Our route for some distance lay along a perfectly level country bearing ample crops of rice. The road was straight and generally bordered with lofty trees forming a fine avenue. It was at first sandy, afterwards grassy, with occasional streams and mudholes. At a distance about four miles we reached Mataram, the capital of the island and the residence of the Rajah. It is a large village with wide streets bordered by a magnificent avenue of trees, and low houses concealed behind mud walls. Within this royal city no native of the lower orders is allowed to ride, and our attendant, a Javanese, was obliged to dismount and lead his horse while

Volume I

we rode slowly through. The abodes of the Rajah and of the High Priest are distinguished by pillars of red brick constructed with much taste; but the palace itself seemed to differ but little from the ordinary houses of the country. Beyond Mataram and close to it is Karangassam, the ancient residence of the native or Sassak Rajahs before the conquest of the island by the Balinese.

Soon after passing Mataram the country began gradually to rise in gentle undulations, swelling occasionally into low hills towards the two mountainous tracts in the northern and southern parts of the island. It was now that I first obtained an adequate idea of one of the most wonderful systems of cultivation in the world, equalling all that is related of Chinese industry, and as far as I know surpassing in the labour that has been bestowed upon it any tract of equal extent in the most civilized countries of Europe. I rode through this strange garden utterly amazed and hardly able to realize the fact that in this remote and little known island, from which all Europeans except a few traders at the port are jealously excluded, many hundreds of square miles of irregularly undulating country have been so skillfully terraced and levelled, and so permeated by artificial channels, that every portion of it can be irrigated and dried at pleasure. According as the slope of the ground is more or less rapid, each terraced plot consists in some places of many acres, in others of a few square yards. We saw them in every state of cultivation; some in stubble, some being ploughed, some with rice-crops in various stages of growth. Here were luxuriant patches of tobacco; there, cucumbers, sweet potatoes, yams, beans or Indian-corn varied the scene. In some places the ditches were dry, in others little streams crossed our road and were distributed over lands about to be sown or planted. The banks which bordered every terrace rose regularly in horizontal lines above each other; sometimes rounding an abrupt knoll and looking like a fortification, or sweeping around some deep hollow and forming on a gigantic scale the seats of an amphitheatre. Every brook and rivulet had been diverted from its bed, and instead of flowing along the lowest ground, were to be found crossing our road half-way up an ascent, yet bordered by ancient trees and moss-grown stones so as to have all the appearance of a natural channel, and bearing testimony to the remote period at which the work had been done. As we advanced further into the country, the scene was diversified by abrupt rocky hills, by steep ravines, and by clumps of bamboos and palm-trees near houses or villages; while in the distance the fine range of mountains of which Lombock Peak, eight thousand feet high, is the culminating point, formed a fit background to a view scarcely to be surpassed either in human interest or picturesque beauty.

Along the first part of our road we passed hundreds of women carrying rice, fruit, and vegetables to market; and further on, an almost uninterrupted line of horses laden with rice in bags or in the ear, on their way to the port of Ampanam. At every few miles along the road, seated under shady trees or slight sheds, were sellers of sugar-cane, palm-wine, cooked rice, salted eggs, and fried plantains, with a few other native delicacies. At these stalls a hearty meal may be made for a penny, but we contented ourselves with drinking some sweet palm-wine, a most delicious beverage in the heat of the day. After having travelled about twenty miles we reached a higher and drier region, where, water being scarce, cultivation was confined to the little flats bordering the streams. Here the country was as beautiful as before, but of a different character; consisting of undulating downs of short turf interspersed with fine clumps of trees and bushes, sometimes the woodland, sometimes the open ground predominating. We only passed through one small patch of true forest, where we were shaded by lofty trees, and saw around us a dark and dense vegetation, highly agreeable after the heat and glare of the open country.

At length, about an hour after noon, we reached our destination—the village of Coupang, situated nearly in the centre of the island—and entered the outer court of a house belonging to one of the chiefs with whom my friend Mr. Ross had a slight acquaintance. Here we were requested to seat ourselves under an open shed with a raised floor of bamboo, a place used to receive visitors and hold audiences.

Turning our horses to graze on the luxuriant grass of the courtyard, we waited until the great man's Malay interpreter appeared, who inquired our business and informed us that the Pumbuckle (chief) was at the Rajah's house, but would soon be back. As we had not yet breakfasted, we begged he would get us something to eat, which he promised to do as soon as possible. It was however about two hours before anything appeared, when a small tray was brought containing two saucers of rice, four small fried fish, and a few vegetables. Having made as good a breakfast as we could, we strolled about the village, and returning, amused ourselves by conversation with a number of men and boys who gathered around us; and by exchanging glances and smiles with a number of women and girls who peeped at us through half-opened doors and other crevices. Two little boys named Mousa and Isa (Moses and Jesus) were great friends with us, and an impudent little rascal called Kachang (a bean) made us all laugh by his mimicry and antics.

At length, about four o'clock, the Pumbuckle made his appearance, and we informed him of our desire to stay with him a few days, to shoot birds and see the country. At this he seemed somewhat disturbed, and asked if we had brought a letter from the Anak Agong (Son of Heaven) which is the title of the Rajah of Lombock. This we had not done, thinking it quite unnecessary; and he then abruptly told us that he must go and speak to his Rajah, to see if we could stay. Hours passed away, night came, and he did not return. I began to think we were suspected of some evil designs, for the Pumbuckle was evidently afraid of getting himself into trouble. He is a Sassak prince, and, though a supporter of the present Rajah, is related to some of the heads of a conspiracy which was quelled a few years since.

About five o'clock a pack-horse bearing my guns and clothes arrived, with my men Ali and Manuel, who had come on foot. The sun set, and it soon became dark, and we got rather hungry as we sat wearily under the shed and no one came. Still hour after hour we waited, until about nine o'clock, the Pumbuckle, the Rajah, some priests, and a number of their followers arrived and took their seats around us. We shook hands, and for some minutes there was a dead silence. Then the Rajah asked what we wanted; to which Mr. Ross replied by endeavouring to make them understand who we were, and why we had come, and that we had no sinister intentions whatever; and that we had not brought a letter from the "Anak Agong," merely because we had thought it quite unnecessary. A long conversation in the Bali language then took place, and questions were asked about my guns, and what powder I had, and whether I used shot or bullets; also what the birds were for, and how I preserved them, and what was done with them in England. Each of my answers and explanations was followed by a low and serious conversation which we could not understand, but the purport of which we could guess. They were evidently quite puzzled, and did not believe a word we had told them. They then inquired if we were really English, and not Dutch; and although we strongly asserted our nationality, they did not seem to believe us.

After about an hour, however, they brought us some supper (which was the same as the breakfast, but without the fish), and after it some very weak coffee and pumpkins boiled with sugar. Having discussed this, a second conference took place; questions were again asked, and the answers again commented on. Between whiles lighter topics were discussed. My spectacles (concave glasses) were tried in succession by three or four old men, who could not make out why they could not see through them, and the fact no doubt was another item of suspicion against me. My beard, too, was the subject of some admiration, and many questions were asked about personal peculiarities which it is not the custom to allude to in European society. At length, about one in the morning, the whole party rose to depart, and, after conversing some time at the gate, all went away. We now begged the interpreter, who with a few boys and men remained about us, to show us a place to sleep in, at which he seemed very much surprised, saying he thought we were very well accommodated where we were. It was quite chilly, and we were very thinly clad and had brought no blankets, but all we could get after another hour's talk was a native mat and pillow, and a few old curtains to hang round three sides of the open shed and protect us a little from the cold breeze. We passed the rest of the night very uncomfortably, and determined to return in the morning and not submit any longer to such shabby treatment.

We rose at daybreak, but it was near an hour before the interpreter made his appearance. We then asked to have some coffee and to see the Pumbuckle, as we wanted a horse for Ali, who was lame, and wished to bid him adieu. The man looked puzzled at such unheard-of demands and vanished into the inner court, locking the door behind him and leaving us again to our meditations. An hour passed and no one came, so I ordered the horses to be saddled and the pack-horse to be loaded, and prepared to start. Just then the interpreter came up on horse back, and looked aghast at our preparations. "Where is the Pumbuckle?" we asked. "Gone to the Rajah's," said he. "We are going," said I. "Oh! pray don't," said he; "wait a little; they are having a consultation, and some priests are coming to see you, and a chief is going off to Mataram to ask the permission of the Anak Agong for you to stay." This settled the matter. More talk, more delay, and another eight or ten hours' consultation were not to be endured; so we started at once, the poor interpreter almost weeping at our obstinacy and hurry, and assuring us "the Pumbuckle would be very sorry, and the Rajah would be very sorry, and if we would but wait all would be right."

I gave Ali my horse, and started on foot, but he afterwards mounted behind Mr. Ross's groom, and we got home very well, though rather hot and tired.

At Mataram we called at the house of Gusti Gadioca, one of the princes of Lombock, who was a friend of Mr. Carter's, and who had promised to show me the guns made by native workmen. Two guns were exhibited, one six, the other seven feet long, and of a proportionably large bore. The barrels were twisted and well finished, though not so finely worked as ours. The stock was well made, and extended to the end of the barrel. Silver and gold ornament was inlaid over most of the surface, but the locks were taken from English muskets. The Gusti assured me, however, that the Rajah had a man who made locks and also rifled barrels. The workshop where these guns are made and the tools used were next shown us, and were very remarkable. An open shed with a couple of small mud forges were the chief objects visible. The bellows consisted of two bamboo cylinders, with pistons worked by hand. They move very easily, having a loose stuffing of feathers thickly set round the piston so as to act as a valve, and produce a regular blast. Both cylinders communicate with the same nozzle, one piston rising while the other falls. An oblong piece of iron on the ground was the anvil, and a small vice was fixed on the projecting root of a tree outside. These, with a few files and hammers, were literally the only tools with which an old man makes these fine guns, finishing then himself from the rough iron and wood.

I was anxious to know how they bored these long barrels, which seemed perfectly true and are said to shoot admirably; and, on asking the Gusti, received the enigmatical answer: "We use a basket full of stones." Being utterly unable to imagine what he could mean, I asked if I could see how they did it, and one of the dozen little boys around us was sent to fetch the basket. He soon returned with this most extraordinary boring-machine, the mode of using which the Gusti then explained to me. It was simply a strong bamboo basket, through the bottom of which was stuck upright a pole about three feet long, kept in its place by a few sticks tied across the top with rattans.

The bottom of the pole has an iron ring, and a hole in which four-cornered borers of hardened iron can be fitted. The barrel to be bored is buried upright in the ground, the borer is inserted into it, the top of the stick or vertical shaft is held by a cross-piece of bamboo with a hole in it, and the basket is filled with stones to get the required weight. Two boys turn the bamboo round. The barrels are made in pieces of about eighteen inches long, which are first bored small, and then welded together upon a straight iron rod. The whole barrel is then worked with borers of gradually increasing size, and in three days the boring is finished. The whole matter was explained in such a straightforward manner that I have no doubt the process described to me was that actually used; although, when examining one of the handsome, well-finished, and serviceable guns, it was very hard to realize the fact that they had been made from first to last with tools hardly sufficient for an English blacksmith to make a horseshoe.

The day after we returned from our excursion, the Rajah came to Ampanam to a feast given by Gusti Gadioca, who resides there; and soon after his arrival we went to have an audience. We found him in a large courtyard sitting on a mat under a shady tree; and all his followers, to the number of three or four hundred, squatting on the ground in a large circle round him. He wore a sarong or Malay petticoat and a green jacket. He was a man about thirty-five years of age, and of a pleasing countenance, with some appearance of intellect combined with indecision. We bowed, and took our seats on the ground near some chiefs we were acquainted with, for while the Rajah sits no one can stand or sit higher. He first inquired who I was, and what I was doing in Lombock, and then requested to see some of my birds. I accordingly sent for one of my boxes of bird-skins and one of insects, which he examined carefully, and seemed much surprised that they could be so well preserved. We then had a little conversation about Europe and the Russian war, in which all natives take an interest. Having heard much of a country-seat of the Rajah's called Gunong Sari, I took the opportunity to ask permission to visit it and shoot a few birds there which he immediately granted. I then thanked him, and we took our leave.

An hour after, his son came to visit Mr. Carter accompanied by about a hundred followers, who all sat on the ground while he came into the open shed where Manuel was skinning birds. After some time he went into the house, had a bed arranged to sleep a little, then drank some wine, and after an hour or two had dinner brought him from the Gusti's house, which he ate with eight of the principal priests and princes, he pronounced a blessing over the rice and commenced eating first, after which the rest fell to. They rolled up balls of rice in their hands, dipped them in the gravy and swallowed them rapidly, with little pieces of meat and fowl cooked in a variety of ways. A boy fanned the young Rajah while eating. He was a youth of about fifteen, and had already three wives. All wore the kris, or Malay crooked dagger, on the beauty and value of which they greatly pride themselves. A companion of the Rajah's had one with a golden handle, in which were set twenty-eight diamonds and several other jewels. He said it had cost him £700. The sheaths are of ornamental wood and ivory, often covered on one side with gold. The blades are beautifully veined with white metal worked into the iron, and they are kept very carefully. Every man without exception carries a kris, stuck behind into the large waist-cloth which all wear, and it is generally the most valuable piece of property he possesses.

A few days afterwards our long-talked-of excursion to Gunong Sari took place. Our party was increased by the captain and supercargo of a Hamburg ship loading with rice for China. We were mounted on a very miscellaneous lot of Lombock ponies, which we had some difficulty in supplying with the necessary saddles, etc.; and most of us had to patch up our girths, bridles, or stirrup-leathers as best we could. We passed through Mataram, where we were joined by our friend Gusti Gadioca, mounted on a handsome black horse, and riding as all the natives do, without saddle or stirrups, using only a handsome saddlecloth and very ornamental bridle.

About three miles further, along pleasant byways, brought us to the place. We entered through a rather handsome brick gateway supported by hideous Hindu deities in stone. Within was an enclosure with two square fish-ponds and some fine trees; then another gateway through which we entered into a park. On the right was a brick house, built somewhat in the Hindu style, and placed on a high terrace or platform; on the left a large fish-pond, supplied by a little rivulet which entered it out of the mouth of a gigantic crocodile well executed in brick and stone. The edges of the pond were bricked, and in the centre rose a fantastic and picturesque pavilion ornamented with grotesque statues. The pond was well stocked with fine fish, which come every morning to be fed at the sound of a wooden gong which is hung near for the purpose. On striking it a number of fish immediately came out of the masses of weed with which the pond abounds, and followed us along the margin expecting food. At the same time some deer came out of as adjacent wood, which, from being seldom shot at and regularly fed, are almost tame. The jungle and woods which surrounded the park appearing to abound in birds, I went to shoot a few, and was rewarded by getting several specimens of the fine new kingfisher, Halcyon fulgidus, and the curious and handsome ground thrush, Zoothera andromeda. The former belies its name by not frequenting water or feeding on fish. It lives constantly in low damp thickets picking up ground insects, centipedes, and small mollusca. Altogether I was much pleased with my visit to this place, and it gave me a higher opinion than I had before entertained of the taste of these people, although the style of the buildings and of the sculpture is very much inferior to those of the magnificent ruins in Java.

I must now say a few words about the character, manners, and customs of these interesting people.

The aborigines of Lombock are termed Sassaks. They are a Malay race hardly differing in appearance from the people of Malacca or Borneo. They are Mahometans and form the bulk of the population. The ruling classes, on the other hand, are natives of the adjacent island of Bali, and are of the Brahminical religion. The government is an absolute monarchy, but it seems to be conducted with more wisdom and moderation than is usual in Malay countries. The father of the present Rajah conquered the island, and the people seem now quite reconciled to their new rulers, who do not interfere with their religion, and probably do not tax them any heavier than did the native chiefs they have supplanted. The laws now in force in Lombock are very severe. Theft is punished by death. Mr. Carter informed me that a man once stole a metal coffee-pot from his house. He was caught, the pot restored, and the man brought to Mr. Carter to punish as he thought fit. All the natives recommended Mr. Carter to have him "krissed" on the spot; "for if you don't," said they, "he will rob you again." Mr. Carter, however, let him off with a warning, that if he ever came inside his premises again he would certainly be shot. A few months afterwards the same man stole a horse from Mr. Carter. The horse was recovered, but the thief was not caught. It is an established rule, that anyone found in a house after dark, unless with the owner's knowledge, may be stabbed, his body thrown out into the street or upon the beach, and no questions will be asked.

The men are exceedingly jealous and very strict with their wives. A married woman may not accept a cigar or a sirih leaf from a stranger under pain of death. I was informed that some years ago one of the English traders had a Balinese woman of good family living with him—the connection being considered quite honourable by the natives. During some festival this girl offended against the law by accepting a flower or some such trifle from another man. This was reported to the Rajah (to some of whose wives the girl was related), and he immediately sent to the Englishman's house ordering him to give the woman up as she must be "krissed." In vain he begged and prayed, and offered to pay any fine the Rajah might impose, and finally refused to give her up unless he was forced to do so. This the Rajah did not wish to resort to, as he no doubt thought he was acting as much for the Englishman's honour as for his own; so he appeared to let the matter drop. But some time afterwards he sent one of his followers to the house, who beckoned the girl to the door, and then saying, "The Rajah sends you this," stabbed her to the heart. More serious infidelity is punished still more cruelly, the woman and her paramour being tied back to back and thrown into the sea, where some large crocodiles are always on the watch to devour the bodies. One such execution took place while I was at Ampanam, but I took a long walk into the country to be out of the way until it was all over, thus missing the opportunity of having a horrible narrative to enliven my somewhat tedious story.

One morning, as we were sitting at breakfast, Mr. Carter's servant informed us that there was an "Amok" in the village—in other words, that a man was "running a muck." Orders were immediately given to shut and fasten the gates of our enclosure; but hearing nothing for some time, we went out, and found there had been a false alarm, owing to a slave having run away, declaring he would "amok," because his master wanted to sell him. A short time before, a man had been killed at a gaming-table because, having lost half-a-dollar more than he possessed, he was going to "amok." Another had killed or wounded seventeen people before he could be destroyed. In their wars a whole regiment of these people will sometimes agree to "amok," and then rush on with such energetic desperation as to be very formidable to men not so excited as themselves. Among the ancients these would have been looked upon as heroes or demigods who sacrificed themselves for their country. Here it is simply said—they made "amok."

Macassar is the most celebrated place in the East for "running a muck." There are said to be one or two a month on the average, and five, ten, or twenty persons are sometimes killed or wounded at one of them. It is the national, and therefore the honourable, mode of committing suicide among the natives of Celebes, and is the fashionable way of escaping from their difficulties. A Roman fell upon his sword, a Japanese rips up his stomach, and an Englishman blows out his brains with a pistol. The Bugis mode has many advantages to one suicidally inclined. A man thinks himself wronged by society—he is in debt and cannot pay—he is taken for a slave or has gambled away his wife or child into slavery—he sees no way of recovering what he has lost, and becomes desperate. He will not put up with such cruel wrongs, but will be revenged on mankind and die like a hero. He grasps his kris-handle, and the next moment draws out the weapon and stabs a man to the heart. He runs on, with bloody kris in his hand, stabbing at everyone he meets. "Amok! Amok!" then resounds through the streets. Spears, krisses, knives and guns are brought out against him. He rushes madly forward, kills all he can—men, women, and children—and dies overwhelmed by numbers amid all the excitement of a battle. And what that excitement is those who have been in one best know, but all who have ever given way to violent passions, or even indulged in violent and exciting exercises, may form a very good idea. It is a delirious intoxication, a temporary madness that absorbs every thought and every energy. And can we wonder at the kris-bearing, untaught, brooding Malay preferring such a death, looked upon as almost honourable to the cold-blooded details of suicide, if he wishes to escape from overwhelming troubles, or the merciless clutches of the hangman and the disgrace of a public execution, when he has taken the law into his own hands and too hastily revenged himself upon his enemy? In either case he chooses rather to "amok."

The great staples of the trade of Lombock as well as of Bali are rice and coffee; the former grown on the plains, the latter on the hills. The rice is exported very largely to other islands of the Archipelago, to Singapore, and even to China, and there are generally one or more vessels loading in the port. It is brought into Ampanam on pack-horses, and almost every day a string of these would come into Mr. Carter's yard. The only money the natives will take for their rice is Chinese copper cash, twelve hundred of which go to a dollar. Every morning two large sacks of this money had to be counted out into convenient sums for payment. From Bali quantities of dried beef and ox-tongues are exported, and from Lombock a good many ducks and ponies. The ducks are a peculiar breed, which have very long flat bodies, and walk erect almost like penguins. They are generally of a pale reddish ash colour, and are kept in large flocks. They are very cheap and are largely consumed by the crews of the rice ships, by whom they are called Baly-soldiers, but are more generally known elsewhere as penguin-ducks.

My Portuguese bird-stuffer Fernandez now insisted on breaking his agreement and returning to Singapore; partly from homesickness, but more I believe from the idea that his life was not worth many months' purchase among such bloodthirsty and uncivilized peoples. It was a considerable loss to me, as I had paid him full three times the usual wages for three months in advance, half of which was occupied in the voyage and the rest in a place where I could have done without him, owing to there being so few insects that I could devote my own time to shooting and skinning. A

Volume I

few days after Fernandez had left, a small schooner came in bound for Macassar, to which place I took a passage. As a fitting conclusion to my sketch of these interesting islands, I will narrate an anecdote which I heard of the present Rajah; and which, whether altogether true or not, well illustrates native character, and will serve as a means of introducing some details of the manners and customs of the country to which I have not yet alluded.

CHAPTER XII. LOMBOCK: HOW THE RAJAH TOOK THE CENSUS.

The Rajah of Lombock was a very wise man and he showed his wisdom greatly in the way he took the census. For my readers must know that the chief revenues of the Rajah were derived from a head-tax of rice, a small measure being paid annually by every man, woman, and child in the island, There was no doubt that every one paid this tax, for it was a very light one, and the land was fertile and the people well off; but it had to pass through many hands before it reached the Government storehouses. When the harvest was over the villagers brought their rice to the Kapala kampong, or head of the village; and no doubt he sometimes had compassion for the poor or sick and passed over their short measure, and sometimes was obliged to grant a favour to those who had complaints against him; and then he must keep up his own dignity by having his granaries better filled than his neighbours, and so the rice that he took to the "Waidono" that was over his district was generally good deal less than it should have been. And all the "Waidonos" had of course to take care of themselves, for they were all in debt and it was so easy to take a little of the Government rice, and there would still be plenty for the Rajah. And the "Gustis" or princes who received the rice from the Waidonos helped themselves likewise, and so when the harvest was all over and the rice tribute was all brought in, the quantity was found to be less each year than the one before. Sickness in one district, and fevers in another, and failure of the crops in a third, were of course alleged as the cause of this falling off; but when the Rajah went to hunt at the foot of the great mountain, or went to visit a "Gusti" on the other side of the island, he always saw the villages full of people, all looking well-fed and happy. And he noticed that the krisses of his chiefs and officers were getting handsomer and handsomer; and the handles that were of yellow wood were changed for ivory, and those of ivory were changed for gold, and diamonds and emeralds sparkled on many of them; and he knew very well which way the tribute-rice went. But as he could not prove it he kept silence, and resolved in his own heart someday to have a census taken, so that he might know the number of his people, and not be cheated out of more rice than was just and reasonable.

But the difficulty was how to get this census. He could not go himself into every village and every house, and count all the people; and if he ordered it to be done by the regular officers they would quickly understand what it was for, and the census would be sure to agree exactly with the quantity of rice he got last year. It was evident therefore that to answer his purpose no one must suspect why the census was taken;

and to make sure of this, no one must know that there was any census taken at all. This was a very hard problem; and the Rajah thought and thought, as hard as a Malay Rajah can be expected to think, but could not solve it; and so he was very unhappy, and did nothing but smoke and chew betel with his favourite wife, and eat scarcely anything; and even when he went to the cock-fight did not seem to care whether his best birds won or lost. For several days he remained in this sad state, and all the court were afraid some evil eye had bewitched the Rajah; and an unfortunate Irish captain who had come in for a cargo of rice and who squinted dreadfully, was very nearly being krissed, but being first brought to the royal presence was graciously ordered to go on board and remain there while his ship stayed in the port.

One morning however, after about a week's continuance of this unaccountable melancholy, a welcome change took place, for the Rajah sent to call together all the chiefs, priests, and princes who were then in Mataram, his capital city; and when they were all assembled in anxious expectation, he thus addressed them:

"For many days my heart has been very sick and I knew not why, but now the trouble is cleared away, for I have had a dream. Last night the spirit of the 'Gunong Agong'—the great fire mountain—appeared to me, and told me that I must go up to the top of the mountain. All of you may come with me to near the top, but then I must go up alone, and the great spirit will again appear to me and will tell me what is of great importance to me and to you and to all the people of the island. Now go all of you and make this known through the island, and let every village furnish men to make clear a road for us to go through the forest and up the great mountain."

So the news was spread over the whole island that the Rajah must go to meet the great spirit on the top of the mountain; and every village sent forth its men, and they cleared away the jungle and made bridges over the mountain streams and smoothed the rough places for the Rajah's passage. And when they came to the steep and craggy rocks of the mountain, they sought out the best paths, sometimes along the bed of a torrent, sometimes along narrow ledges of the black rocks; in one place cutting down a tall tree so as to bridge across a chasm, in another constructing ladders to mount the smooth face of a precipice. The chiefs who superintended the work fixed upon the length of each day's journey beforehand according to the nature of the road, and chose pleasant places by the banks of clear streams and in the neighbourhood of shady trees, where they built sheds and huts of bamboo well thatched with the leaves of palm-trees, in which the Rajah and his attendants might eat and sleep at the close of each day.

And when all was ready, the princes and priests and chief men came again to the Rajah, to tell him what had been done and to ask him when he would go up the mountain. And he fixed a day, and ordered every man of rank and authority to accompany him, to do honour to the great spirit who had bid him undertake the journey, and to show how willingly they obeyed his commands. And then there was much preparation throughout the whole island. The best cattle were killed and the meat salted and sun-dried; and abundance of red peppers and sweet potatoes were gathered; and the tall pinang-trees were climbed for the spicy betel nut, the sirih-leaf was tied up in bundles, and every man filled his tobacco pouch and lime box to the brim, so that he might not want any of the materials for chewing the refreshing betel during the journey. The stores of provisions were sent on a day in advance. And on the day before that appointed for starting, all the chiefs both great and small came to Mataram, the abode of the king, with their horses and their servants, and the bearers of their sirih boxes, and their sleeping-mats, and their provisions. And they encamped under the tall Waringin-trees that border all the roads about Mataram, and with blazing fires frighted away the ghouls and evil spirits that nightly haunt the gloomy avenues.

In the morning a great procession was formed to conduct the Rajah to the mountain. And the royal princes and relations of the Rajah mounted their black horses whose tails swept the ground; they used no saddle or stirrups, but sat upon a cloth of gay colours; the bits were of silver and the bridles of many-coloured cords. The less important people were on small strong horses of various colours, well suited to a mountain journey; and all (even the Rajah) were bare-legged to above the knee, wearing only the gay coloured cotton waist-cloth, a silk or cotton jacket, and a large handkerchief tastefully folded around the head. Everyone was attended by one or two servants bearing his sirih and betel boxes, who were also mounted on ponies; and great numbers more had gone on in advance or waited to bring up the rear. The men in authority were numbered by hundreds and their followers by thousands, and all the island wondered what great thing would come of it.

For the first two days they went along good roads and through many villages which were swept clean, and where bright cloths were hung out at the windows; and all the people, when the Rajah came, squatted down upon the ground in respect, and every man riding got off his horse and squatted down also, and many joined the procession at every village. At the place where they stopped for the night, the people had placed stakes along each side of the roads in front of the houses. These were split crosswise at the top, and in the cleft were fastened little clay lamps, and between them were stuck the green leaves of palm-trees, which, dripping with the evening dew, gleamed prettily with the many twinkling lights. And few went to sleep that night until the morning hours, for every house held a knot of eager talkers, and much betel-nut was consumed, and endless were the conjectures what would come of it.

On the second day they left the last village behind them and entered the wild country that surrounds the great mountain, and rested in the huts that had been prepared for them on the banks of a stream of cold and sparkling water. And the Rajah's hunters, armed with long and heavy guns, went in search of deer and wild bulls in the surrounding woods, and brought home the meat of both in the early morning, and sent it on in advance to prepare the midday meal. On the third day they advanced as far as horses could go, and encamped at the foot of high rocks, among which narrow pathways only could be found to reach the mountain-top. And on the fourth morning when the Rajah set out, he was accompanied only by a small party of priests and princes with their immediate attendants; and they toiled wearily up the rugged way, and sometimes were carried by their servants, until they passed up above the great trees, and then among the thorny bushes, and above them again on to the black and burned rock of the highest part of the mountain.

And when they were near the summit, the Rajah ordered them all to halt, while he alone went to meet the great spirit on the very peak of the mountain. So he went on with two boys only who carried his sirih and betel, and soon reached the top of the mountain among great rocks, on the edge of the great gulf whence issue forth continually smoke and vapour. And the Rajah asked for sirih, and told the boys to sit down under a rock and look down the mountain, and not to move until he returned to them. And as they were tired, and the sun was warm and pleasant, and the rock sheltered them from the cold wind, the boys fell asleep. And the Rajah went a little way on under another rock; and as he was tired, and the sun was warm and pleasant, and he too fell asleep.

And those who were waiting for the Rajah thought him a long time on the top of the mountain, and thought the great spirit must have much to say, or might perhaps want to keep him on the mountain always, or perhaps he had missed his way in coming down again. And they were debating whether they should go and search for him, when they saw him coming down with the two boys. And when he met them he looked very grave, but said nothing; and then all descended together, and the procession returned as it had come; and the Rajah went to his palace and the chiefs to their villages, and the people to their houses, to tell their wives and children all that had happened, and to wonder yet again what would come of it.

And three days afterwards the Rajah summoned the priests and the princes and the chief men of Mataram, to hear what the great spirit had told him on the top of the mountain. And when they were all assembled, and the betel and sirih had been handed round, he told them what had happened. On the top of the mountain he had fallen into a trance, and the great spirit had appeared to him with a face like burnished gold, and had said—"Oh Rajah! much plague and sickness and fevers are coming upon all the earth, upon men and upon horses and upon cattle; but as you and your people have obeyed me and have come up to my great mountain, I will teach you how you and all the people of Lombock may escape this plague." And all waited anxiously, to hear how they were to be saved from so fearful a calamity. And after a short silence the Rajah spoke again and told them, that the great spirit had commanded that twelve sacred krisses should be made, and that to make them every village and every district must send a bundle of needles—a needle for every head in the village. And when any grievous disease appeared in any village, one of the sacred krisses should be sent there; and if every house in that village had sent the right number of needles, the disease would immediately cease; but if the number of needles sent had not been exact, the kris would have no virtue.

So the princes and chiefs sent to all their villages and communicated the wonderful news; and all made haste to collect the needles with the greatest accuracy, for they feared that if but one were wanting, the whole village would suffer. So one by one the head men of the villages brought in their bundles of needles; those who were near Mataram came first, and those who were far off came last; and the Rajah received them with his own hands and put them away carefully in an inner chamber, in a camphor-wood chest whose hinges and clasps were of silver; and on every bundle was marked the name of the village and the district from whence it came, so that it might be known that all had heard and obeyed the commands of the great spirit.

And when it was quite certain that every village had sent in its bundle, the Rajah divided the needles into twelve equal parts, and ordered the best steelworker in Mataram to bring his forge and his bellows and his hammers to the palace, and to make the twelve krisses under the Rajah's eye, and in the sight of all men who chose to see it. And when they were finished, they were wrapped up in new silk and put away carefully until they might be wanted.

Now the journey to the mountain was in the time of the east wind when no rain falls in Lombock. And soon after the krisses were made it was the time of the rice harvest, and the chiefs of districts and of villages brought their tax to the Rajah according to the number of heads in their villages. And to those that wanted but little of the full amount, the Rajah said nothing; but when those came who brought only half or a fourth part of what was strictly due, he said to them mildly, "The needles which you sent from your village were many more than came from such-a-one's village, yet your tribute is less than his; go back and see who it is that has not paid the tax." And the next year the produce of the tax increased greatly, for they feared that the Rajah might justly kill those who a second time kept back the right tribute. And so the Rajah became very rich, and increased the number of his soldiers, and gave golden jewels to his wives, and bought fine black horses from the white-skinned Hollanders, and made great feasts when his children were born or were married; and none of the Rajahs or Sultans among the Malays were so great or powerful as the Rajah of Lombock.

And the twelve sacred krisses had great virtue. And, when any sickness appeared in a village one of them was sent for; and sometimes the sickness went away, and then the sacred kris was taken back again with great Honour, and the head men of the village came to tell the Rajah of its miraculous power, and to thank him. And sometimes the sickness would not go away; and then everybody was convinced that there had been a mistake in the number of needles sent from that village, and therefore the sacred kris had no effect, and had to be taken back again by the head men with heavy hearts, but still, with all honour—for was not the fault their own?

CHAPTER XIII. TIMOR.

(COUPANG, 1857-1869. DELLI, 1861.)

THE island of Timor is about three hundred miles long and sixty wide, and seems to form the termination of the great range of volcanic islands which begins with Sumatra more than two thousand miles to the west. It differs however very remarkably from all the other islands of the chain in not possessing any active volcanoes, with the one exception of Timor Peak near the centre of the island, which was formerly active, but was blown up during an eruption in 1638 and has since been quiescent. In no other part of Timor do there appear to be any recent igneous rocks, so that it can hardly be classed as a volcanic island. Indeed its position is just outside of the great volcanic belt, which extends from Flores through Ombay and Wetter to Banda.

I first visited Timor in 1857, staying a day at Coupang, the chief Dutch town at the west end of the island; and again in May 1859, when I stayed a fortnight in the same neighbourhood. In the spring of 1861 I spent four months at Delli, the capital of the Portuguese possessions in the eastern part of the island.

The whole neighbourhood of Coupang appears to have

been elevated at a recent epoch, consisting of a rugged surface of coral rock, which rises in a vertical wall between the beach and the town, whose low, white, red-tiled houses give it an appearance very similar to other Dutch settlements in the East. The vegetation is everywhere scanty and scrubby. Plants of the families Apocynaceae and Euphorbiaceae, abound; but there is nothing that can be called a forest, and the whole country has a parched and desolate appearance, contrasting strongly with the lofty forest trees and perennial verdure of the Moluccas or of Singapore. The most conspicuous feature of the vegetation was the abundance of fine fan-leaved palms (Borassus flabelliformis), from the leaves of which are constructed the strong and durable water-buckets in general use, and which are much superior to those formed from any other species of palm. From the same tree, palm-wine and sugar are made, and the common thatch for houses formed of the leaves lasts six or seven years without removal. Close to the town I noticed the foundation of a ruined house below high-water mark, indicating recent subsidence. Earthquakes are not severe here, and are so infrequent and harmless that the chief houses are built of stone.

The inhabitants of Coupang consist of Malays, Chinese, and Dutch, besides the natives, so that there are many strange and complicated mixtures among the population. There is one resident English merchant, and whalers as well as Australian ships often come here for stores and water. The native Timorese preponderate, and a very little examination serves to show that they have nothing in common with Malays, but are much more closely allied to the true Papuans of the Aru Islands and New Guinea. They are tall, have pronounced features, large somewhat aquiline noses, and frizzly hair, and are generally of a dusky brown colour. The way in which the women talk to each other and to the men, their loud voices and laughter, and general character of self-assertion, would enable an experienced observer to decide, even without seeing them, that they were not Malays.

Mr. Arndt, a German and the Government doctor, invited me to stay at his house while in Coupang, and I gladly accepted his offer, as I only intended making a short visit. We at first began speaking French, but he got on so badly that we soon passed insensibly into Malay; and we afterwards held long discussions on literary, scientific, and philosophical questions in that semi-barbarous language, whose deficiencies we made up by the free use of French or Latin words.

After a few walks in the neighbourhood of the town, I found such a poverty of insects and birds that I determined to go for a few days to the island of Semao at the western extremity of Timor, where I heard that there was forest country with birds not found at Coupang. With some difficulty I obtained a large dugout boat with outriggers, to take me over a distance of about twenty miles. I found the country pretty well wooded, but covered with shrubs and thorny bushes rather than forest trees, and everywhere excessively parched and dried up by the long-continued dry season. I stayed at the village of Oeassa, remarkable for its soap springs. One of these is in the middle of the village, bubbling out from a little cone of mud to which the ground rises all round like a volcano in miniature. The water has a soapy feel and produces a strong lather when any greasy substance is washed in it. It contains alkali and iodine, in such quantities as to destroy all vegetation for some distance around. Close by the village is one of the finest springs I have ever seen, contained in several rocky basins communicating by narrow channels. These have been neatly walled where required and partly levelled, and form fine natural baths. The water is well tasted and clear as crystal, and the basins are surrounded by a grove of lofty many-stemmed banyan-trees, which keep them always cool and shady, and add greatly to the picturesque beauty of the scene.

The village consists of curious little houses very different from any I have seen elsewhere. They are of an oval figure, and the walls are made of sticks about four feet high placed close together. From this rises a high conical roof thatched with grass. The only opening is a door about three feet high. The people are like the Timorese with frizzly or wavy hair and of a coppery brown colour. The better class appear to have a mixture of some superior race which has much improved their features. I saw in Coupang some chiefs from the island of Savu further west, who presented characters very distinct from either the Malay or Papuan races. They most resembled Hindus, having fine well-formed features and straight thin noses with clear brown complexions. As the Brahminical religion once spread over all Java, and even now exists in Bali and Lombock, it is not at all improbable that some natives of India should have reached this island, either by accident or to escape persecution, and formed a permanent settlement there.

I stayed at Oeassa four days, when, not finding any insects and very few new birds, I returned to Coupang to await the next mail steamer. On the way I had a narrow escape of being swamped. The deep coffin-like boat was filled up with my baggage, and with vegetables, cocoa-nut and other fruit for Coupang market, and when we had got some way across into a rather rough sea, we found that a quantity of water was coming in which we had no means of baling out. This caused us to sink deeper in the water, and then we shipped seas over our sides, and the rowers, who had before declared it was nothing, now became alarmed and turned the boat round to get back to the coast of Semao, which was not far off. By clearing away some of the baggage a little of the water could be baled out, but hardly so fast as it came in, and when we neared the coast we found nothing but vertical walls of rock against which the sea was violently beating. We coasted along some distance until we found a little cove, into which we ran the boat, hauled it on shore, and emptying it found a large hole in the bottom, which had been temporarily stopped up with a plug of cocoa-nut which had come out. Had we been a quarter of a mile further off before we discovered the leak, we should certainly have been obliged to throw most of our baggage overboard, and might easily have lost our lives. After we had put all straight and secure we again started, and when we were halfway across got into such a strong current and high cross sea that we were very nearly being swamped a second time, which made me vow never to trust myself again in such small and miserable vessels.

The mail steamer did not arrive for a week, and I occupied myself in getting as many of the birds as I could, and found some which were very interesting. Among them were five species of pigeons of as many distinct genera, and most of them peculiar to the island; two parrots—the fine red-winged broad-tail (Platycercus vulneratus), allied to an Australian species, and a green species of the genus Geoffroyus. The Tropidorhynchus timorensis was as ubiquitous and as noisy as I had found it at Lombock; and the Sphaecothera viridis, a curious green oriole with bare red orbits, was a great acquisition. There were several pretty finches, warblers, and flycatchers, and among them I obtained the elegant blue and red Cyornis hyacinthina; but I cannot recognise among my collections the species mentioned by Dampier, who seems to have been much struck by the number of small songbirds in Timor. He says: "One sort of these pretty little birds my men called the ringing bird, because it had six notes, and always repeated all his notes twice, one after the other, beginning high and shrill and ending low. The bird was about the bigness of a lark, having a small, sharp, black bill and blue wings; the head and breast were of a pale red, and there was a blue streak about its neck." In Semao, monkeys are abundant. They are the common hare-lipped monkey (Macacus cynomolgus), which is found all over the western islands of the Archipelago, and may have been introduced by natives, who often carry it about captive. There are also some deer, but it is not quite certain whether they are of the same species as are found in Java.

I arrived at Delli, the capital of the Portuguese possessions in Timor, on January 12, 1861, and was kindly received by Captain Hart, an Englishman and an old resident, who trades in the produce of the country and cultivates coffee on an estate at the foot of the hills. With him I was introduced to Mr. Geach, a mining-engineer who had been for two years endeavouring to discover copper in sufficient quantity to be worth working.

Delli is a most miserable place compared with even the poorest of the Dutch towns. The houses are all of mud and thatch; the fort is only a mud enclosure; and the custom-house and church are built of the same mean materials, with no attempt at decoration or even neatness. The whole aspect of the place is that of a poor native town, and there is no sign of cultivation or civilization round about it. His Excellency the Governor's house is the only one that makes any pretensions to appearance, and that is merely a low whitewashed cottage or bungalow. Yet there is one thing in which civilization exhibits itself—officials in black and white European costume, and officers in gorgeous uniforms abound in a degree quite disproportionate to the size or appearance of the place.

The town being surrounded for some distance by swamps and mudflats is very unhealthy, and a single night often gives a fever to newcomers which not unfrequently proves fatal. To avoid this malaria, Captain Hart always slept at his plantation, on a slight elevation about two miles from the town, where Mr. Geach also had a small house, which he kindly invited me to share. We rode there in the evening; and in the course of two days my baggage was brought up, and I was able to look about me and see if I could do any collecting.

For the first few weeks I was very unwell and could not go far from the house. The country was covered with low spiny shrubs and acacias, except in a little valley where a stream came down from the hills, where some fine trees and bushes shaded the water and formed a very pleasant place to ramble up. There were plenty of birds about, and of a tolerable variety of species; but very few of them were gaily coloured. Indeed, with one or two exceptions, the birds of this tropical island were hardly so ornamental as those of Great Britain. Beetles were so scarce that a collector might fairly say there were none, as the few obscure or uninteresting species would not repay him for the search. The only insects at all remarkable or interesting were the butterflies, which, though comparatively few in species, were sufficiently abundant, and comprised a large proportion of new or rare sorts. The banks of the stream formed my best collecting-ground, and I daily wandered up and down its shady bed, which about a mile up became rocky and precipitous. Here I obtained the rare and beautiful swallow-tail butterflies, Papilio aenomaus and P. liris; the males of which are quite unlike each other, and belong in fact to distinct sections of the genus, while the females are so much alike that they are undistinguishable on the wing, and to an uneducated eye equally so in the cabinet. Several other beautiful butterflies rewarded my search in this place, among which I may especially mention the Cethosia leschenaultii, whose wings of the deepest purple are bordered with buff in such a manner as to resemble at first sight our own Camberwell beauty, although it belongs to a different genus. The most abundant butterflies were the whites and yellows (Pieridae), several of which I had already found at Lombock and at Coupang, while others were new to me.

Early in February we made arrangements to stay for a week at a village called Baliba, situated about four miles off on the mountains, at an elevation of 2,000 feet. We took our baggage and a supply of all necessaries on packhorses; and though the distance by the route we took was not more than six or seven miles, we were half a day getting there. The roads were mere tracks, sometimes up steep rocky stairs, sometimes in narrow gullies worn by the horses' feet, and where it was necessary to tuck up our legs on our horses' necks to avoid having them crushed. At some of these places the baggage had to be unloaded, at others it was knocked off. Sometimes the ascent or descent was so steep that it was easier to walk than to cling to our ponies' backs; and thus we went up and down over bare hills whose surface was covered with small pebbles and scattered over with Eucalypti, reminding me of what I had read of parts of the interior of Australia rather than of the Malay Archipelago.

The village consisted of three houses only, with low walls raised a few feet on posts, and very high roofs thatched with grass hanging down to within two or three feet of the ground. A house which was unfinished and partly open at the back was given for our use, and in it we rigged up a table, some benches, and a screen, while an inner enclosed portion served us for a sleeping apartment. We had a splendid view down upon Delli and the sea beyond. The country around was undulating and open, except in the hollows, where there were some patches of forest, which Mr. Geach, who had been all over the eastern part of Timor, assured me was the most luxuriant he had yet seen in the island. I was in hopes of finding some insects here, but was much disappointed, owing perhaps to the dampness of the climate; for it was not until the sun was pretty high that the mists cleared away, and by noon we were generally clouded up again, so that there was seldom more than an hour or two of fitful sunshine. We searched in every direction for birds and other game, but they were very scarce. On our way I had shot the fine white-headed pigeon, Ptilonopus cinctus, and the pretty little lorikeet, Trichoglossus euteles. I got a few more of these at the blossoms of the Eucalypti, and also the allied species Trichoglossus iris, and a few other small but interesting birds. The common jungle-cock of India (Gallus bankiva) was found here, and furnished us with some excellent meals; but we could get no deer. Potatoes are grown higher up the mountains in abundance, and are very good. We had a sheep killed every other day, and ate our mutton with much appetite in the cool climate, which rendered a fire always agreeable.

Although one-half the European residents in Delli are continually ill from fever, and the Portuguese have occupied the place for three centuries, no one has yet built a house on these fine hills, which, if a tolerable road were made, would be only an hour's ride from the town; and almost equally good situations might be found on a lower level at half an hour's distance. The fact that potatoes and wheat of excellent quality are grown in abundance at from 3,000 to 3,500 feet elevation, shows what the climate and soil are capable of if properly cultivated. From one to two thousand feet high, coffee would thrive; and there are hundreds of square miles of country over which all the varied products which require climates between those of coffee and wheat would flourish; but no attempt has yet been made to form a single mile of road, or a single acre of plantation!

There must be something very unusual in the climate of Timor to permit wheat being grown at so moderate an elevation. The grain is of excellent quality, the bread made from it being equal to any I have ever tasted, and it is universally acknowledged to be unsurpassed by any made from imported European or American flour. The fact that the natives have (quite of their own accord) taken to cultivating such foreign articles as wheat and potatoes, which they bring in small quantities on the backs of ponies by the most horrible mountain tracks, and sell very cheaply at the seaside, sufficiently indicates what might be done if good roads were made, and if the people were taught, encouraged, and protected. Sheep also do well on the mountains; and a breed of hardy ponies in much repute all over the Archipelago, runs half-wild, so that it appears as if this island, so barren-looking and devoid of the usual features of tropical vegetation, were yet especially adapted to supply a variety of products essential to Europeans, which the other islands will not produce, and which they accordingly import from the other side of the globe.

On the 24th of February my friend Mr. Geach left Timor, having finally reported that no minerals worth working were to be found. The Portuguese were very much annoyed, having made up their minds that copper is abundant, and still believing it to be so. It appears that from time immemorial pure native copper has been found at a place on the coast about thirty miles east of Delli.

The natives say they find it in the bed of a ravine, and many years ago a captain of a vessel is said to have got some hundreds-weight of it. Now, however, it is evidently very scarce, as during the two years Mr. Geach resided in the country, none was found. I was shown one piece several pounds' weight, having much the appearance of one of the larger Australian nuggets, but of pure copper instead of gold. The natives and the Portuguese have very naturally imagined that where these fragments come from there must be more; and they have a report or tradition, that a mountain at the head of the ravine is almost pure copper, and of course of immense value.

After much difficulty a company was at length formed to work the copper mountain, a Portuguese merchant of Singapore supplying most of the capital. So confident were they of the existence of the copper, that they thought it would be waste of time and money to have any exploration made first; and accordingly, sent to England for a mining engineer, who was to bring out all necessary tools, machinery, laboratory, utensils, a number of mechanics, and stores of all kinds for two years, in order to commence work on a copper-mine which he was told was already discovered. On reaching Singapore a ship was freighted to take the men and stores to Timor, where they at length arrived after much delay, a long voyage, and very great expense.

A day was then fixed to "open the mines." Captain Hart accompanied Mr. Geach as interpreter. The Governor, the Commandante, the Judge, and all the chief people of the place went in state to the mountain, with Mr. Geach's assistant and some of the workmen. As they went up the valley Mr. Geach examined the rocks, but saw no signs of copper. They went on and on, but still nothing except a few mere traces of very poor ore. At length they stood on the copper mountain itself. The Governor stopped, the officials formed a circle, and he then addressed them, saying, that at length the day had arrived they had all been so long expecting, when the treasures of the soil of Timor would be brought to light, and much more in very grandiloquent Portuguese; and concluded by turning to Mr. Geach, and requesting him to point out the best spot for them to begin work at once, and uncover the mass of virgin copper. As the ravines and precipices among which they had passed, and which had been carefully examined, revealed very clearly the nature and mineral constitution of the country, Mr. Geach simply told them that there was not a trace of copper there, and that it was perfectly useless to begin work. The audience were thunderstruck! The Governor could not believe his ears. At length, when Mr. Geach had repeated his statement, the Governor told him severely that he was mistaken; that they all knew there was copper there in abundance, and all they wanted him to tell them, as a mining-engineer, was how best to get at it; and that at all events he was to begin work somewhere. This Mr. Geach refused to do, trying to explain that the ravines had cut far deeper into the hill than he could do in years, and that he would not throw away money or time on any such useless attempt. After this speech had been interpreted to him, the Governor saw it was no use, and without saying a word turned his horse and rode away, leaving my friends alone on the mountain. They all believed there was some conspiracy that the Englishman would not find the copper, and that they had been cruelly betrayed.

Mr. Geach then wrote to the Singapore merchant who was his employer, and it was arranged that he should send the mechanics home again, and himself explore the country for minerals. At first the Government threw obstacles in his way and entirely prevented his moving; but at length he was allowed to travel about, and for more than a year he and his assistant explored the eastern part of Timor, crossing it in several places from sea to sea, and ascending every important valley, without finding any minerals that would pay the expense of working. Copper ore exists in several places, but always too poor in quality. The best would pay well if situated in England; but in the interior of an utterly barren country, with roads to make, and all skilled labour and materials to import, it would have been a losing concern. Gold also occurs, but very sparingly and of poor quality. A fine spring of pure petroleum was discovered far in the interior, where it can never be available until the country is civilized. The whole affair was a dreadful disappointment to the Portuguese Government, who had considered it such a certain thing that they had contracted for the Dutch mail steamers to stop at Delli and several vessels from Australia were induced to come with miscellaneous cargoes, for which they expected to find a ready sale among the population at the newly-opened mines. The lumps of native copper are still, however, a mystery. Mr. Geach has examined the country in every direction without being able to trace their origin; so that it seems probable that they result from the debris of old copper-bearing strata, and are not really more abundant than gold nuggets are in Australia or California. A high reward was offered to any native who should find a piece and show the exact spot where he obtained it, but without effect.

Volume I

The mountaineers of Timor are a people of Papuan type, having rather slender forms, bushy frizzled hair, and the skin of a dusky brown colour. They have the long nose with overhanging apex which is so characteristic of the Papuan, and so absolutely unknown among races of Malayan origin. On the coast there has been much admixture of some of the Malay races, and perhaps of Hindu, as well as of Portuguese. The general stature there is lower, the hair wavy instead of frizzled, and the features less prominent. The houses are built on the ground, while the mountaineers raise theirs on posts three or four feet high. The common dress is a long cloth, twisted around the waist and hanging to the knee, as shown in the illustration (page 305), copied from a photograph. Both men carry the national umbrella, made of an entire fan-shaped palm leaf, carefully stitched at the fold of each leaflet to prevent splitting. This is opened out, and held sloping over the head and back during a shower. The small water-bucket is made from an entire unopened leaf of the same palm, and the covered bamboo probably contains honey for sale. A curious wallet is generally carried, consisting of a square of strongly woven cloth, the four corners of which are connected by cords, and often much ornamented with beads and tassels. Leaning against the house behind the figure on the right are bamboos, used instead of water jars.

A prevalent custom is the "pomali," exactly equivalent to the "taboo" of the Pacific islanders, and equally respected. It is used on the commonest occasions, and a few palm leaves stuck outside a garden as a sign of the "pomali" will preserve its produce from thieves as effectually as the threatening notice of man-traps, spring guns, or a savage dog would do with us. The dead are placed on a stage, raised six or eight feet above the ground, sometimes open and sometimes covered with a roof. Here the body remains until the relatives can afford to make a feast, when it is buried. The Timorese are generally great thieves, but are not bloodthirsty. They fight continually among themselves, and take every opportunity of kidnapping unprotected people of other tribes for slaves; but Europeans may pass anywhere through the country in safety. Except for a few half-breeds in the town, there are no native Christians in the island of Timor. The people retain their independence in a great measure, and both dislike and despise their would-be rulers, whether Portuguese or Dutch.

The Portuguese government in Timor is a most miserable one. Nobody seems to care the least about the improvement of the country, and at this time, after three hundred years of occupation, there has not been a mile of road made beyond the town, and there is not a solitary European resident anywhere in the interior. All the Government officials oppress and rob the natives as much as they can, and yet there is no care taken to render the town defensible should the Timorese attempt to attack it. So ignorant are the military officers, that having received a small mortar and some shells, no one could be found who knew how to use them; and during an insurrection of the natives (while I was at Delli) the officer who expected to be sent against the insurgents was instantly taken ill! And they were allowed to get possession of an important pass within three miles of the town, where they could defend themselves against ten times the force. The result was that no provisions were brought down from the hills; a famine was imminent; and the Governor had to send off to beg for supplies from the Dutch Governor of Amboyna.

In its present state Timor is more trouble than profit to its Dutch and Portuguese rulers, and it will continue to be so unless a different system is pursued. A few good roads into the elevated districts of the interior; a conciliatory policy and strict justice towards the natives, and the introduction of a good system of cultivation as in Java and northern Celebes, might yet make Timor a productive and valuable island. Rice grows well on the marshy flats, which often fringe the coast, and maize thrives in all the lowlands, and is the common food of the natives as it was when Dampier visited the island in 1699. The small quantity of coffee now grown is of very superior quality, and it might be increased to any extent. Sheep thrive, and would always be valuable as fresh food for whalers and to supply the adjacent islands with mutton, if not for their wool; although it is probable that on the mountains this product might soon be obtained by judicious breeding. Horses thrive amazingly; and enough wheat might be grown to supply the whole Archipelago if there were sufficient inducements to the natives to extend its cultivation, and good roads by which it could be cheaply transported to the coast.

Under such a system the natives would soon perceive that European government was advantageous to them. They would begin to save money, and property being rendered secure they would rapidly acquire new wants and new tastes, and become large consumers of European goods. This would be a far surer source of profit to their rulers than imposts and extortion, and would be at the same time more likely to produce peace and obedience than the mock-military rule which has hitherto proved most ineffective. To inaugurate such a system would however require an immediate outlay of capital, which neither Dutch nor Portuguese seem inclined to make, and a number of honest and energetic officials, which the latter nation at least seems unable to produce; so that it is much to be feared that Timor will for many years to come remain in its present state of chronic insurrection and misgovernment.

Morality at Delli is at as low an ebb as in the far interior of Brazil, and crimes are connived at which would entail infamy and criminal prosecution in Europe. While I was there it was generally asserted and believed in the place, that two officers had poisoned the husbands of women with whom they were carrying on intrigues, and with whom they immediately cohabited on the death of their rivals. Yet no one ever thought for a moment of showing disapprobation of the crime, or even of considering it a crime at all, the husbands in question being low half-castes, who of course ought to make way for the pleasures of their superiors.

Judging from what I saw myself and by the descriptions of Mr. Geach, the indigenous vegetation of Timor is poor and monotonous. The lower ranges of the hills are everywhere covered with scrubby Eucalypti, which only occasionally grow into lofty forest trees. Mingled with these in smaller quantities are acacias and the fragrant sandalwood, while the higher mountains, which rise to about six or seven thousand feet, are either covered with coarse grass or are altogether barren. In the lower grounds are a variety of weedy bushes, and open waste places are covered everywhere with a nettle-like wild mint. Here is found the beautiful crown lily, Gloriosa superba, winding among the bushes, and displaying its magnificent blossoms in great profusion. A wild vine also occurs, bearing great irregular bunches of hairy grapes of a coarse but very luscious flavour. In some of the valleys where the vegetation is richer, thorny shrubs and climbers are so abundant as to make the thickets quite impenetrable.

The soil seems very poor, consisting chiefly of decomposing clayey shales; and the bare earth and rock is almost everywhere visible. The drought of the hot season is so severe that most of the streams dry up in the plains before they reach the sea; everything becomes burned up, and the leaves of the larger trees fall as completely as in our winter. On the mountains from two to four thousand feet elevation there is a much moister atmosphere, so that potatoes and other European products can be grown all the year round. Besides ponies, almost the only exports of Timor are sandalwood and beeswax. The sandalwood (Santalum sp.) is the produce of a small tree, which grows sparingly in the mountains of Timor and many of the other islands in the far East. The wood is of a fine yellow colour, and possesses a well-known delightful fragrance which is wonderfully permanent. It is brought down to Delli in small logs, and is chiefly exported to China, where it is largely used to burn in the temples, and in the houses of the wealthy.

The beeswax is a still more important and valuable product, formed by the wild bees (Apis dorsata), which build huge honeycombs, suspended in the open air from the underside of the lofty branches of the highest trees. These are of a semicircular form, and often three or four feet in diameter. I once saw the natives take a bees' nest, and a very interesting sight it was. In the valley where I used to collect insects, I one day saw three or four Timorese men and boys under a high tree, and, looking up, saw on a very lofty horizontal branch three large bees' combs. The tree was straight and smooth-barked and without a branch, until at seventy or eighty feet from the ground it gave out the limb which the bees had chosen for their home. As the men were evidently looking after the bees, I waited to watch their operations. One of them first produced a long piece of wood apparently the stem of a small tree or creeper, which he had brought with him, and began splitting it through in several directions, which showed that it was very tough and stringy. He then wrapped it in palm-leaves, which were secured by twisting a slender creeper round them. He then fastened his cloth tightly round his loins, and producing another cloth wrapped it around his head, neck, and body, and tied it firmly around his neck, leaving his face, arms, and legs completely bare. Slung to his girdle he carried a long thin coil of cord; and while he had been making these preparations, one of his companions had cut a strong creeper or bush-rope eight or ten yards long, to one end of which the wood-torch was fastened, and lighted at the bottom, emitting a steady stream of smoke. Just above the torch a chopping-knife was fastened by a short cord.

The bee-hunter now took hold of the bush-rope just above the torch and passed the other end around the trunk of the tree, holding one end in each hand. Jerking it up the tree a little above his head he set his foot against the trunk, and leaning back began walking up it. It was wonderful to see the skill with which he took advantage of the slightest irregularities of the bark or obliquity of the stem to aid his ascent, jerking the stiff creeper a few feet higher when he had found a firm hold for his bare foot. It almost made me giddy to look at him as he rapidly got up—thirty, forty, fifty feet above the ground; and I kept wondering how he could possibly mount the next few feet of straight smooth trunk. Still, however, he kept on with as much coolness and apparent certainty as if he were going up a ladder, until he got within ten or fifteen feet of the bees. Then he stopped a moment, and took care to swing the torch (which hung just at his feet) a little towards these dangerous insects, so as to send up the stream of smoke between him and them. Still going on, in a minute more he brought himself under the limb, and, in a manner quite unintelligible to me, seeing that both hands were occupied in supporting himself by the creeper, managed to get upon it.

By this time the bees began to be alarmed, and formed a dense buzzing swarm just over him, but he brought the torch up closer to him, and coolly brushed away those that settled on his arms or legs. Then stretching himself along the limb, he crept towards the nearest comb and swung the torch just under it. The moment the smoke touched it, its colour changed in a most curious manner from black to white, the myriads of bees that had covered it flying off and forming a dense cloud above and around. The man then lay at full length along the limb, and brushed off the remaining bees with his hand, and then drawing his knife cut off the comb at one slice close to the tree, and attaching the thin cord to it, let it down to his companions below. He was all this time enveloped in a crowd of angry bees, and how he bore their stings so coolly, and went on with his work at that giddy height so deliberately, was more than I could understand. The bees were evidently not stupified by the smoke or driven away far by it, and it was impossible that the small stream from the torch could protect his whole body when at work. There were three other combs on the same tree, and all were successively taken, and furnished the whole party with a luscious feast of honey and young bees, as well as a valuable lot of wax.

After two of the combs had been let down, the bees became rather numerous below, flying about wildly and stinging viciously. Several got about me, and I was soon stung, and had to run away, beating them off with my net and capturing them for specimens. Several of them followed me for at least half a mile, getting into my hair and persecuting me most pertinaciously, so that I was more astonished than ever at the immunity of the natives. I am inclined to think that slow and deliberate motion, and no attempt at escape, are perhaps the best safeguards. A bee settling on a passive native probably behaves as it would on a tree or other inanimate substance, which it does not attempt to sting. Still they must often suffer, but they are

The Malay Archipelago

used to the pain and learn to bear it impassively, as without doing so no man could be a bee-hunter.

CHAPTER XIV. THE NATURAL HISTORY OF THE TIMOR GROUP.

IF we look at a map of the Archipelago, nothing seems more unlikely than that the closely connected chain of islands from Java to Timor should differ materially in their natural productions. There are, it is true, certain differences of climate and of physical geography, but these do not correspond with the division the naturalist is obliged to make. Between the two ends of the chain there is a great contrast of climate, the west being exceedingly moist and leaving only a short and irregular dry season, the east being as dry and parched up, and having but a short wet season. This change, however, occurs about the middle of Java, the eastern portion of that island having as strongly marked seasons as Lombock and Timor. There is also a difference in physical geography; but this occurs at the eastern termination of the chain where the volcanoes which are the marked feature of Java, Bali, Lombock, Sumbawa, and Flores, turn northwards through Gunong Api to Banda, leaving Timor with only one volcanic peak near its centre, while the main portion of the island consists of old sedimentary rocks. Neither of these physical differences corresponds with the remarkable change in natural productions which occurs at the Straits of Lombock, separating the island of that name from Bali, and which is at once so large in amount and of so fundamental a character, as to form an important feature in the zoological geography of our globe.

The Dutch naturalist Zollinger, who resided a long time on the island of Bali, informs us that its productions completely assimilate with those of Java, and that he is not aware of a single animal found in it which does not inhabit the larger island. During the few days which I stayed on the north coast of Bali on my way to Lombock, I saw several birds highly characteristic of Javan ornithology. Among these were the yellow-headed weaver (Ploceus hypoxantha), the black grasshopper thrush (Copsychus amoenus), the rosy barbet (Megalaema rosea), the Malay oriole (Oriolus horsfieldi), the Java ground starling (Sturnopastor jalla), and the Javanese three-toed woodpecker (Chrysonotus tiga). On crossing over to Lombock, separated from Bali by a strait less than twenty miles wide, I naturally expected to meet with some of these birds again; but during a stay there of three months I never saw one of them, but found a totally different set of species, most of which were utterly unknown not only in Java, but also in Borneo, Sumatra, and Malacca. For example, among the commonest birds in Lombock were white cockatoos and three species of Meliphagidae or honeysuckers, belonging to family groups which are entirely absent from the western or Indo-Malayan region of the

Archipelago. On passing to Flores and Timor the distinctness from the Javanese productions increases, and we find that these islands form a natural group, whose birds are related to those of Java and Australia, but are quite distinct from either. Besides my own collections in Lombock and Timor, my assistant Mr. Allen made a good collection in Flores; and these, with a few species obtained by the Dutch naturalists, enable us to form a very good idea of the natural history of this group of islands, and to derive therefrom some very interesting results.

The number of birds known from these islands up to this date is: 63 from Lombock, 86 from Flores, and 118 from Timor; and from the whole group, 188 species. With the exception of two or three species which appear to have been derived from the Moluccas, all these birds can be traced, either directly or by close allies, to Java on the one side or to Australia on the other; although no less than 82 of them are found nowhere out of this small group of islands. There is not, however, a single genus peculiar to the group, or even one which is largely represented in it by peculiar species; and this is a fact which indicates that the fauna is strictly derivative, and that its origin does not go back beyond one of the most recent geological epochs. Of course there are a large number of species (such as most of the waders, many of the raptorial birds, some of the kingfishers, swallows, and a few others), which range so widely over a large part of the Archipelago that it is impossible to trace them as having come from any one part rather than from another. There are fifty-seven such species in my list, and besides these there are thirty-five more which, though peculiar to the Timor group, are yet allied to wide-ranging forms. Deducting these ninety-two species, we have nearly a hundred birds left whose relations with those of other countries we will now consider.

If we first take those species which, as far as we yet know, are absolutely confined to each island, we find, in:

Lombock 4 belonging to 2 genera, of which 1 is Australian, 1 Indian.
Flores   12   "   7   "   5 are   "   2   "
Timor   42   "   20   "   16 are   "   4   "

The actual number of peculiar species in each island I do not suppose to be at all accurately determined, since the rapidly increasing numbers evidently depend upon the more extensive collections made in Timor than in Flores, and in Flores than in Lombock; but what we can depend more upon, and what is of more special interest, is the greatly increased proportion of Australian forms and decreased proportion of Indian forms, as we go from west to east. We shall show this in a yet more striking manner by counting the number of species identical with those of Java and Australia respectively in each island, thus:

Volume I

	In Lombock.	In Flores.	In Timor.
Javan birds	33	23	11
Australian birds	4	5	10

Here we see plainly the course of the migration which has been going on for hundreds or thousands of years, and is still going on at the present day. Birds entering from Java are most numerous in the island nearest Java; each strait of the sea to be crossed to reach another island offers an obstacle, and thus a smaller number get over to the next island. [The names of all the birds inhabiting these islands are to be found in the "Proceedings of the Zoological Society of London" for the year 1863.] It will be observed that the number of birds that appear to have entered from Australia is much less than those which have come from Java; and we may at first sight suppose that this is due to the wide sea that separates Australia from Timor. But this would be a hasty and, as we shall soon see, an unwarranted supposition. Besides these birds identical with species inhabiting Java and Australia, there are a considerable number of others very closely allied to species peculiar to those countries, and we must take these also into account before we form any conclusion on the matter. It will be as well to combine these with the former table thus:

	In Lombock.	In Flores.	In Timor.
Javan birds	33	23	11
Closely allied to Javan birds.	1	5	6
Total	34	28	17
Australian birds	4	5	10
Closely allied to Australian birds	3	9	26
Total.....	7	14	36

We now see that the total number of birds which seem to have been derived from Java and Australia is very nearly equal, but there is this remarkable difference between the two series: that whereas the larger proportion by far of the Java set are identical with those still inhabiting that country, an almost equally large proportion of the Australian set are distinct, though often very closely allied species. It is to be observed also, that these representative or allied species diminish in number as they recede from Australia, while they increase in number as they recede from Java. There are two reasons for this, one being that the islands decrease rapidly in size from Timor to Lombock, and can therefore support a decreasing number of species; the other and the more important is, that the distance of Australia from Timor cuts off the supply of fresh immigrants, and has thus allowed variation to have full play; while the vicinity of Lombock to Bali and Java has allowed a continual influx of fresh individuals which, by crossing with the earlier immigrants, has checked variation.

To simplify our view of the derivative origin of the birds of these islands let us treat them as a whole, and thus perhaps render more intelligible their respective relations to Java and Australia.

The Timor group of islands contains:

Javan birds....... 36 Australian birds... 13 Closely allied species.. 11 Closely allied species.. 35 Derived from Java .... 47 Derived from Australia... 48

We have here a wonderful agreement in the number of birds belonging to Australian and Javanese groups, but they are divided in exactly a reverse manner, three-fourths of the Javan birds being identical species and one-fourth representatives, while only one-fourth of the Australian forms are identical and three-fourths representatives. This is the most important fact which we can elicit from a study of the birds of these islands, since it gives us a very complete clue to much of their past history.

Change of species is a slow process—on that we are all agreed, though we may differ about how it has taken place. The fact that the Australian species in these islands have mostly changed, while the Javan species have almost all remained unchanged, would therefore indicate that the district was first peopled from Australia. But, for this to have been the case, the physical conditions must have been very different from what they are now. Nearly three hundred miles of open sea now separate Australia from Timor, which island is connected with Java by a chain of broken land divided by straits which are nowhere more than about twenty miles wide. Evidently there are now great facilities for the natural productions of Java to spread over and occupy the whole of these islands, while those of Australia would find very great difficulty in getting across. To account for the present state of things, we should naturally suppose that Australia was once much more closely connected with Timor than it is at present; and that this was the case is rendered highly probable by the fact of a submarine bank extending along all the north and west coast of Australia, and at one place approaching within twenty miles of the coast of Timor. This indicates a recent subsidence of North Australia, which probably once extended as far as the edge of this bank, between which and Timor there is an unfathomed depth of ocean.

I do not think that Timor was ever actually connected with Australia, because such a large number of very abundant and characteristic groups of Australian birds are quite absent, and not a single Australian mammal has entered Timor—which would certainly not have been the case had the lands been actually united. Such groups as the bower birds (Ptilonorhynchus), the black and red cockatoos (Calyptorhynchus), the blue wrens (Malurus), the crowshrikes (Cracticus), the Australian shrikes (Falcunculus and Colluricincla), and many others, which abound all over Australia, would certainly have spread into Timor if it had been united to that country, or even if for any long time it had approached nearer to it than twenty miles. Neither do any of the most characteristic groups of Australian insects occur in Timor; so that everything combines to indicate that a strait of the sea has always separated it from Australia, but that at one period this strait was reduced to a width of about twenty miles.

But at the time when this narrowing of the sea took place in one direction, there must have been a greater separation at the other end of the chain, or we should find more equality in the numbers of identical and representative species derived from each extremity. It is true that the widening of the strait at the Australian end by subsidence, would, by putting a stop to immigration and intercrossing of individuals from the mother country, have allowed full scope to the causes which have led to the modification of the species; while the continued stream of immigrants from Java, would, by continual intercrossing, check such modification. This view will not, however, explain all the facts; for the character of the fauna of the Timorese group is indicated as well by the forms which are absent from it as by those which it contains, and is by this kind of evidence shown to be much more Australian than Indian. No less than twenty-nine genera, all more or less abundant in Java, and most of which range over a wide area, are altogether absent; while of the equally diffused Australian genera only about fourteen are wanting. This would clearly indicate that there has been, until recently, a wide separation from Java; and the fact that the islands of Bali and Lombock are small, and are almost wholly volcanic, and contain a smaller number of modified forms than the other islands, would point them out as of comparatively recent origin. A wide arm of the sea probably occupied their place at the time when Timor was in the closest proximity to Australia; and as the subterranean fires were slowly piling up the now fertile islands of Bali and Lombock, the northern shores of Australia would be sinking beneath the ocean. Some such changes as have been here indicated, enable us to understand how it happens, that though the birds of this group are on the whole almost as much Indian as Australian, yet the species which are peculiar to the group are mostly Australian in character; and also why such a large number of common Indian forms which extend through Java to Bali, should not have transmitted a single representative to the island further east.

The Mammalia of Timor as well as those of the other

islands of the group are exceedingly scanty, with the exception of bats. These last are tolerably abundant, and no doubt many more remain to be discovered. Out of fifteen species known from Timor, nine are found also in Java, or the islands west of it; three are Moluccan species, most of which are also found in Australia, and the rest are peculiar to Timor.

The land mammals are only seven in number, as follows: 1. The common monkey, Macacus cynomolgus, which is found in all the Indo-Malayan islands, and has spread from Java through Bali and Lombock to Timor. This species is very frequent on the banks of rivers, and may have been conveyed from island to island on trees carried down by floods. 2. Paradoxurus fasciatus; a civet cat, very common over a large part of the Archipelago. 3. Felis megalotis; a tiger cat, said to be peculiar to Timor, where it exists only in the interior, and is very rare. Its nearest allies are in Java. 4. Cervus timoriensis; a deer, closely allied to the Javan and Moluccan species, if distinct. 5. A wild pig, Sus timoriensis; perhaps the same as some of the Moluccan species. 6. A shrew mouse, Sorex tenuis; supposed to be peculiar to Timor. 7. An Eastern opossum, Cuscus orientalis; found also in the Moluccas, if not a distinct species.

The fact that not one of these species is Australian or nearly allied to any Australian form, is strongly corroborative of the opinion that Timor has never formed a part of that country; as in that case some kangaroo or other marsupial animal would almost certainly be found there. It is no doubt very difficult to account for the presence of some of the few mammals that do exist in Timor, especially the tiger cat and the deer. We must consider, however, that during thousands, and perhaps hundreds of thousands of years, these islands and the seas between them have been subjected to volcanic action. The land has been raised and has sunk again; the straits have been narrowed or widened; many of the islands may have been joined and dissevered again; violent floods have again and again devastated the mountains and plains, carrying out to sea hundreds of forest trees, as has often happened during volcanic eruptions in Java; and it does not seem improbable that once in a thousand, or ten thousand years, there should have occurred such a favourable combination of circumstances as would lead to the migration of two or three land animals from one island to another. This is all that we need ask to account for the very scanty and fragmentary group of Mammalia which now inhabit the large island of Timor. The deer may very probably have been introduced by man, for the Malays often keep tame fawns; and it may not require a thousand, or even five hundred years, to establish new characters in an animal removed to a country so different in climate and vegetation as is Timor from the Moluccas. I have not mentioned horses, which are often thought to be wild in Timor, because there are no grounds whatever for such a belief. The Timor ponies have every one an owner, and are quite as much domesticated animals as the cattle on a South American hacienda.

I have dwelt at some length upon the origin of the Timorese fauna because it appears to be a most interesting and instructive problem. It is very seldom that we can trace the animals of a district so clearly as we can in this case to two definite sources, and still more rarely that they furnish such decisive evidence of the time, the manner, and the proportions of their introduction. We have here a group of Oceanic Islands in miniature—islands which have never formed part of the adjacent lands, although so closely approaching them; and their productions have the characteristics of true Oceanic Islands slightly modified. These characteristics are: the absence all Mammalia except bats; and the occurrence of peculiar species of birds, insects, and land shells, which, though found nowhere else, are plainly related to those of the nearest land. Thus, we have an entire absence of all Australian mammals, and the presence of only a few stragglers from the west which can be accounted for in the manner already indicated. Bats are tolerably abundant.

Birds have many peculiar species, with a decided relationship to those of the two nearest masses of land. The insects have similar relations with the birds. As an example, four species of the Papilionidae are peculiar to Timor, three others are also found in Java, and one in Australia. Of the four peculiar species two are decided modifications of Javanese forms, while the others seem allied to those of the Moluccas and Celebes. The very few land shells known are all, curiously enough, allied to or identical with Moluccan or Celebes forms. The Pieridae (white and yellow butterflies) which wander more, and from frequenting open grounds, are more liable to be blown out to sea, seem about equally related to those of Java, Australia, and the Moluccas.

It has been objected to in Mr. Darwin's theory, of Oceanic Islands having never been connected with the mainland, that this would imply that their animal population was a matter of chance; it has been termed the "flotsam and jetsam theory," and it has been maintained that nature does not work by the "CHAPTER of accidents." But in the case which I have here described, we have the most positive evidence that such has been the mode of peopling the islands. Their productions are of that miscellaneous character which we should expect from such an origin; and to suppose that they have been portions of Australia or of Java will introduce perfectly gratuitous difficulties, and render it quite impossible to explain those curious relations which the best known group of animals (the birds) have been shown to exhibit. On the other hand, the depth of the surrounding seas, the form of the submerged banks, and the volcanic character of most of the islands, all point to an independent origin.

Before concluding, I must make one remark to avoid

misapprehension. When I say that Timor has never formed part of Australia, I refer only to recent geological epochs. In Secondary or even Eocene or Miocene times, Timor and Australia may have been connected; but if so, all record of such a union has been lost by subsequent submergence, and in accounting for the present land-inhabitants of any country we have only to consider those changes which have occurred since its last elevation above the waters. Since such last elevation, I feel confident that Timor has not formed part of Australia.

CHAPTER XV. CELEBES.

(MACASSAR, SEPTEMBER TO NOVEMBER, 1856.)

I LEFT Lombock on the 30th of August, and reached Macassar in three days. It was with great satisfaction that I stepped on a shore which I had been vainly trying to reach since February, and where I expected to meet with so much that was new and interesting.

The coast of this part of Celebes is low and flat, lined with trees and villages so as to conceal the interior, except at occasional openings which show a wide extent of bare and marshy rice-fields. A few hills of no great height were visible in the background; but owing to the perpetual haze over the land at this time of the year, I could nowhere discern the high central range of the peninsula, or the celebrated peak of Bontyne at its southern extremity. In the roadstead of Macassar there was a fine 42-gun frigate, the guardship of the place, as well as a small war steamer and three or four little cutters used for cruising after the pirates which infest these seas. There were also a few square-rigged trading-vessels, and twenty or thirty native praus of various sizes. I brought letters of introduction to a Dutch gentleman, Mr. Mesman, and also to a Danish shopkeeper, who could both speak English and who promised to assist me in finding a place to stay, suitable for my pursuits. In the meantime, I went to a kind of clubhouse, in default of any hotel in the place.

Macassar was the first Dutch town I had visited, and I found it prettier and cleaner than any I had yet seen in the East. The Dutch have some admirable local regulations. All European houses must be kept well white-washed, and every person must, at four in the afternoon, water the road in front of his house. The streets are kept clear of refuse, and covered drains carry away all impurities into large open sewers, into which the tide is admitted at high-water and allowed to flow out when it has ebbed, carrying all the sewage with it into the sea. The town consists chiefly of one long narrow street along the seaside, devoted to business, and principally occupied by the Dutch and Chinese merchants' offices and warehouses, and the native shops or bazaars. This extends northwards for more than a mile, gradually merging into native houses often of a most miserable description, but made to have a neat appearance by being all built up exactly to the straight line of the street, and being generally backed by fruit trees. This street is usually thronged with a native population of Bugis and Macassar men, who wear cotton trousers about twelve inches long, covering only from the hip to half-way down the thigh, and the universal Malay sarong, of gay checked colours, worn around the waist or across the shoulders in a variety of ways. Parallel to this street run two short ones which form the old Dutch town, and are enclosed by gates. These consist of private houses, and at their southern end is the fort, the church, and a road at right angles to the beach, containing the houses of the Governor and of the principal officials. Beyond the fort, again along the beach, is another long street of native huts and many country-houses of the tradesmen and merchants. All around extend the flat rice-fields, now bare and dry and forbidding, covered with dusty stubble and weeds. A few months back these were a mass of verdure, and their barren appearance at this season offered a striking contrast to the perpetual crops on the same kind of country in Lombock and Bali, where the seasons are exactly similar, but where an elaborate system of irrigation produces the effect of a perpetual spring.

The day after my arrival I paid a visit of ceremony to the Governor, accompanied by my friend the Danish merchant, who spoke excellent English. His Excellency was very polite, and offered me every facility for travelling about the country and prosecuting my researches in natural history. We conversed in French, which all Dutch officials speak very well.

Finding it very inconvenient and expensive to stay in the town, I removed at the end of a week to a little bamboo house, kindly offered me by Mr. Mesman. It was situated about two miles away, on a small coffee plantation and farm, and about a mile beyond Mr. M.'s own country-house. It consisted of two rooms raised about seven feet above the ground, the lower part being partly open (and serving excellently to skin birds in) and partly used as a granary for rice. There was a kitchen and other outhouses, and several cottages nearby, occupied by men in Mr. M.'s employ.

After being settled a few days in my new house, I found that no collections could be made without going much further into the country. The rice-fields for some miles around resembled English stubbles late in autumn, and were almost as unproductive of bird or insect life. There were several native villages scattered about, so embosomed in fruit trees that at a distance they looked like clumps or patches of forest. These were my only collecting places; but they produced a very limited number of species, and were soon exhausted. Before I could move to any more promising district it was necessary to obtain permission from the Rajah of Goa, whose territories approach to within two miles of the town of Macassar. I therefore presented myself at the Governor's office and requested a letter to the Rajah, to claim his protection, and permission to travel in his territories whenever I might wish to do so. This was immediately granted, and a special messenger was sent with me to carry the letter.

My friend Mr. Mesman kindly lent me a horse, and accompanied me on my visit to the Rajah, with whom he was great friends. We found his Majesty seated out of doors, watching the erection of a new house. He was naked from the waist up, wearing only the usual short trousers and sarong. Two chairs were brought out for us, but all the chiefs and other natives were seated on the ground. The messenger, squatting down at the Rajah's feet, produced the letter, which was sewn up in a covering of yellow silk. It was handed to one of the chief officers, who ripped it open and returned it to the Rajah, who read it, and then showed it to Mr. M., who both speaks and reads the Macassar language fluently, and who explained fully what I required. Permission was immediately granted me to go where I liked in the territories of Goa, but the Rajah desired, that should I wish to stay any time at a place I would first give him notice, in order that he might send someone to see that no injury was done me. Some wine was then brought us, and afterwards some detestable coffee and wretched sweetmeats, for it is a fact that I have never tasted good coffee where people grow it themselves.

Although this was the height of the dry season, and there was a fine wind all day, it was by no means a healthy time of year. My boy Ali had hardly been a day on shore when he was attacked by fever, which put me to great inconvenience, as at the house where I was staying, nothing could be obtained but at mealtime. After having cured Ali, and with much difficulty got another servant to cook for me,

Volume I

I was no sooner settled at my country abode than the latter was attacked with the same disease; and, having a wife in the town, left me. Hardly was he gone than I fell ill myself with strong intermittent fever every other day. In about a week I got over it, by a liberal use of quinine, when scarcely was I on my legs than Ali again became worse than ever. Ali's fever attacked him daily, but early in the morning he was pretty well, and then managed to cook enough for me for the day. In a week I cured him, and also succeeded in getting another boy who could cook and shoot, and had no objection to go into the interior. His name was Baderoon, and as he was unmarried and had been used to a roving life, having been several voyages to North Australia to catch trepang or "beche de mer", I was in hopes of being able to keep him. I also got hold of a little impudent rascal of twelve or fourteen, who could speak some Malay, to carry my gun or insect-net and make himself generally useful. Ali had by this time become a pretty good bird-skinner, so that I was fairly supplied with servants.

I made many excursions into the country, in search of a good station for collecting birds and insects. Some of the villages a few miles inland are scattered about in woody ground which has once been virgin forest, but of which the constituent trees have been for the most part replaced by fruit trees, and particularly by the large palm, Arenga saccharifera, from which wine and sugar are made, and which also produces a coarse black fibre used for cordage. That necessary of life, the bamboo, has also been abundantly planted. In such places I found a good many birds, among which were the fine cream-coloured pigeon, Carpophaga luctuosa, and the rare blue-headed roller, Coracias temmincki, which has a most discordant voice, and generally goes in pairs, flying from tree to tree, and exhibiting while at rest that all-in-a-heap appearance and jerking motion of the head and tail which are so characteristic of the great Fissirostral group to which it belongs. From this habit alone, the kingfishers, bee-eaters, rollers, trogons, and South American puff-birds, might be grouped together by a person who had observed them in a state of nature, but who had never had an opportunity of examining their form and structure in detail. Thousands of crows, rather smaller than our rook, keep up a constant cawing in these plantations; the curious wood-swallows (Artami), which closely resemble swallows in their habits and flight but differ much in form and structure, twitter from the tree-tops; while a lyre-tailed drongo-shrike, with brilliant black plumage and milk-white eyes, continually deceives the naturalist by the variety of its unmelodious notes.

In the more shady parts butterflies were tolerably abundant; the most common being species of Euplaea and Danais, which frequent gardens and shrubberies, and owing to their weak flight are easily captured. A beautiful pale blue and black butterfly, which flutters along near the ground among the thickets, and settles occasionally upon flowers, was one of the most striking; and scarcely less so, was one with a rich orange band on a blackish ground—these both belong to the Pieridae, the group that contains our common white butterflies, although differing so much from them in appearance. Both were quite new to European naturalists. [The former has been named Eronia tritaea; the latter Tachyris ithonae.] Now and then I extended my walks some miles further, to the only patch of true forest I could find, accompanied by my two boys with guns and insect-net. We used to start early, taking our breakfast with us, and eating it wherever we could find shade and water. At such times my Macassar boys would put a minute fragment of rice and meat or fish on a leaf, and lay it on a stone or stump as an offering to the deity of the spot; for though nominal Mahometans the Macassar people retain many pagan superstitions, and are but lax in their religious observances. Pork, it is true, they hold in abhorrence, but will not refuse wine when offered them, and consume immense quantities of "sagueir," or palm-wine, which is about as intoxicating as ordinary beer or cider. When well made it is a very refreshing drink, and we often took a draught at some of the little sheds dignified by the name of bazaars, which are scattered about the country wherever there is any traffic.

One day Mr. Mesman told me of a larger piece of forest

where he sometimes went to shoot deer, but he assured me it was much further off, and that there were no birds. However, I resolved to explore it, and the next morning at five o'clock we started, carrying our breakfast and some other provisions with us, and intending to stay the night at a house on the borders of the wood. To my surprise two hours' hard walking brought us to this house, where we obtained permission to pass the night. We then walked on, Ali and Baderoon with a gun each, Baso carrying our provisions and my insect-box, while I took only my net and collecting-bottle and determined to devote myself wholly to the insects. Scarcely had I entered the forest when I found some beautiful little green and gold speckled weevils allied to the genus Pachyrhynchus, a group which is almost confined to the Philippine Islands, and is quite unknown in Borneo, Java, or Malacca. The road was shady and apparently much trodden by horses and cattle, and I quickly obtained some butterflies I had not before met with. Soon a couple of reports were heard, and coming up to my boys I found they had shot two specimens of one of the finest of known cuckoos, Phoenicophaus callirhynchus. This bird derives its name from its large bill being coloured of a brilliant yellow, red, and black, in about equal proportions. The tail is exceedingly long, and of a fine metallic purple, while the plumage of the body is light coffee brown. It is one of the characteristic birds of the island of Celebes, to which it is confined.

Volume I

After sauntering along for a couple of hours we reached a small river, so deep that horses could only cross it by swimming, so we had to turn back; but as we were getting hungry, and the water of the almost stagnant river was too muddy to drink, we went towards a house a few hundred yards off. In the plantation we saw a small raised hut, which we thought would do well for us to breakfast in, so I entered, and found inside a young woman with an infant. She handed me a jug of water, but looked very much frightened. However, I sat down on the doorstep, and asked for the provisions. In handing them up, Baderoon saw the infant, and started back as if he had seen a serpent. It then immediately struck me that this was a hut in which, as among the Dyaks of Borneo and many other savage tribes, the women are secluded for some time after the birth of their child, and that we did very wrong to enter it; so we walked off and asked permission to eat our breakfast in the family mansion close at hand, which was of course granted. While I ate, three men, two women, and four children watched every motion, and never took eyes off me until I had finished.

On our way back in the heat of the day, I had the good fortune to capture three specimens of a fine Ornithoptera, the largest, the most perfect, and the most beautiful of butterflies. I trembled with excitement as I took the first out of my net and found it to be in perfect condition. The ground colour of this superb insect was a rich shining bronzy black, the lower wings delicately grained with white, and bordered by a row of large spots of the most brilliant satiny yellow. The body was marked with shaded spots of white, yellow, and fiery orange, while the head and thorax were intense black. On the under-side the lower wings were satiny white, with the marginal spots half black and half yellow. I gazed upon my prize with extreme interest, as I at first thought it was quite a new species. It proved however to be a variety of Ornithoptera remus, one of the rarest and most remarkable species of this highly esteemed group. I also obtained several other new and pretty butterflies. When we arrived at our lodging-house, being particularly anxious about my insect treasures, I suspended the box from a bamboo on which I could detect no sign of ants, and then began skinning some of my birds. During my work I often glanced at my precious box to see that no intruders had arrived, until after a longer spell of work than usual I looked again, and saw to my horror that a column of small red ants were descending the string and entering the box. They were already busy at work at the bodies of my treasures, and another half-hour would have seen my whole day's collection destroyed. As it was, I had to take every insect out, clean them thoroughly as well as the box, and then seek a place of safety for them. As the only effectual one, I begged a plate and a basin from my host, filled the former with water, and standing the latter in it placed my box on the top, and then felt secure for the night;

a few inches of clean water or oil being the only barrier these terrible pests are not able to pass.

On returning home to Mamajam (as my house was called) I had a slight return of intermittent fever, which kept me some days indoors. As soon as I was well, I again went to Goa, accompanied by Mr. Mesman, to beg the Rajah's assistance in getting a small house built for me near the forest. We found him at a cock-fight in a shed near his palace, which however, he immediately left to receive us, and walked with us up an inclined plane of boards which serves for stairs to his house. This was large, well-built, and lofty, with bamboo floor and glass windows. The greater part of it seemed to be one large hall divided by the supporting posts. Near a window sat the Queen, squatting on a rough wooden arm-chair, chewing the everlasting sirih and betel-nut, while a brass spittoon by her side and a sirih-box in front were ready to administer to her wants. The Rajah seated himself opposite to her in a similar chair, and a similar spittoon and sirih-box were held by a little boy squatting at his side. Two other chairs were brought for us. Several young women, some the Rajah's daughters, others slaves, were standing about; a few were working at frames making sarongs, but most of them were idle.

And here I might (if I followed the example of most travellers) launch out into a glowing description of the charms of these damsels, the elegant costumes they wore, and the gold and silver ornaments with which they were adorned. The jacket or body of purple gauze would figure well in such a description, allowing the heaving bosom to be seen beneath it, while "sparkling eyes," and "jetty tresses," and "tiny feet" might be thrown in profusely. But, alas! regard for truth will not permit me to expatiate too admiringly on such topics, determined as I am to give as far as I can a true picture of the people and places I visit. The princesses were, it is true, sufficiently good-looking, yet neither their persons nor their garments had that appearance of freshness and cleanliness without which no other charms can be contemplated with pleasure. Everything had a dingy and faded appearance, very disagreeable and unroyal to a European eye. The only thing that excited some degree of admiration was the quiet and dignified manner of the Rajah and the great respect always paid to him. None can stand erect in his presence, and when he sits on a chair, all present (Europeans of course excepted) squat upon the ground. The highest seat is literally, with these people, the place of honour and the sign of rank. So unbending are the rules in this respect, that when an English carriage which the Rajah of Lombock had sent for arrived, it was found impossible to use it because the driver's seat was the highest, and it had to be kept as a show in its coach house. On being told the object of my visit, the Rajah at once said that he would order a house to be emptied for me, which would be much better than building one, as that would take a good deal of time. Bad coffee and sweetmeats were given us as before.

Two days afterwards, I called on the Rajah to ask him to send a guide with me to show me the house I was to occupy. He immediately ordered a man to be sent for, gave him instructions, and in a few minutes we were on our way. My conductor could speak no Malay, so we walked on in silence for an hour, when we turned into a pretty good house and I was asked to sit down. The head man of the district lived here, and in about half an hour we started again, and another hour's walk brought us to the village and where I was to be lodged. We went to the residence of the village chief, who conversed with my conductor for some time.

Getting tired, I asked to be shown the house that was prepared for me, but the only reply I could get was, "Wait a little," and the parties went on talking as before. So I told them I could not wait, as I wanted to see the house and then to go shooting in the forest. This seemed to puzzle them, and at length, in answer to questions, very poorly explained by one or two bystanders who knew a little Malay, it came out that no house was ready, and no one seemed to have the least idea where to get one. As I did not want to trouble the Rajah any more, I thought it best to try to frighten them a little; so I told them that if they did not immediately find me a house as the Rajah had ordered, I should go back and complain to him, but that if a house was found me I would pay for the use of it. This had the desired effect, and one of the head men of the village asked me to go with him and look for a house. He showed me one or two of the most miserable and ruinous description, which I at once rejected, saying, "I must have a good one, and near to the forest." The next he showed me suited very well, so I told him to see that it was emptied the next day, for that the day after I should come and occupy it.

On the day mentioned, as I was not quite ready to go, I sent my two Macassar boys with brooms to sweep out the house thoroughly. They returned in the evening and told me that when they got there the house was inhabited, and not a single article removed. However, on hearing they had come to clean and take possession, the occupants made a move, but with a good deal of grumbling, which made me feel rather uneasy as to how the people generally might take my intrusion into their village. The next morning we took our baggage on three packhorses, and, after a few break-downs, arrived about noon at our destination.

After getting all my things set straight, and having made a hasty meal, I determined if possible to make friends with the people. I therefore sent for the owner of the house and as many of his acquaintances as liked to come, to have a "bitchara," or talk. When they were all seated, I gave them a little tobacco all around, and having my boy Baderoon for interpreter, tried to explain to them why I came there; that

I was very sorry to turn them out of the house, but that the Rajah had ordered it rather than build a new one, which was what I had asked for, and then placed five silver rupees in the owner's hand as one month's rent. I then assured them that my being there would be a benefit to them, as I should buy their eggs and fowls and fruit; and if their children would bring me shells and insects, of which I showed them specimens, they also might earn a good many coppers. After all this had been fully explained to them, with a long talk and discussion between every sentence, I could see that I had made a favourable impression; and that very afternoon, as if to test my promise to buy even miserable little snail-shells, a dozen children came one after another, bringing me a few specimens each of a small Helix, for which they duly received "coppers," and went away amazed but rejoicing.

A few days' exploration made me well acquainted with the surrounding country. I was a long way from the road in the forest which I had first visited, and for some distance around my house were old clearings and cottages. I found a few good butterflies, but beetles were very scarce, and even rotten timber and newly-felled trees (generally so productive) here produced scarcely anything. This convinced me that there was not a sufficient extent of forest in the neighbourhood to make the place worth staying at long, but it was too late now to think of going further, as in about a month the wet season would begin; so I resolved to stay here and get what was to be had. Unfortunately, after a few days I became ill with a low fever which produced excessive lassitude and disinclination to all exertion. In vain I endeavoured to shake it off; all I could do was to stroll quietly each day for an hour about the gardens near, and to the well, where some good insects were occasionally to be found; and the rest of the day to wait quietly at home, and receive what beetles and shells my little corps of collectors brought me daily. I imputed my illness chiefly to the water, which was procured from shallow wells, around which there was almost always a stagnant puddle in which the buffaloes wallowed. Close to my house was an enclosed mudhole where three buffaloes were shut up every night, and the effluvia from which freely entered through the open bamboo floor. My Malay boy Ali was affected with the same illness, and as he was my chief bird-skinner I got on but slowly with my collections.

The occupations and mode of life of the villagers differed but little from those of all other Malay races. The time of the women was almost wholly occupied in pounding and cleaning rice for daily use, in bringing home firewood and water, and in cleaning, dyeing, spinning, and weaving the native cotton into sarongs. The weaving is done in the simplest kind of frame stretched on the floor; and is a very slow and tedious process. To form the checked pattern in common use, each patch of coloured threads has to be pulled up separately by hand and the shuttle passed between them;

so that about an inch a day is the usual progress in stuff a yard and a half wide. The men cultivate a little sirih (the pungent pepper leaf used for chewing with betel-nut) and a few vegetables; and once a year rudely plough a small patch of ground with their buffaloes and plant rice, which then requires little attention until harvest time. Now and then they have to see to the repairs of their houses, and make mats, baskets, or other domestic utensils, but a large part of their time is passed in idleness.

Not a single person in the village could speak more than a few words of Malay, and hardly any of the people appeared to have seen a European before. One most disagreeable result of this was that I excited terror alike in man and beast. Wherever I went, dogs barked, children screamed, women ran away, and men stared as though I were some strange and terrible cannibal or monster. Even the pack-horses on the roads and paths would start aside when I appeared and rush into the jungle; and as to those horrid, ugly brutes, the buffaloes, they could never be approached by me; not for fear of my own but of others' safety. They would first stick out their necks and stare at me, and then on a nearer view break loose from their halters or tethers, and rush away helter-skelter as if a demon were after them, without any regard for what might be in their way. Whenever I met buffaloes carrying packs along a pathway, or being driven home to the village, I had to turn aside into the jungle and hide myself until they had passed, to avoid a catastrophe which would increase the dislike with which I was already regarded. Everyday about noon the buffaloes were brought into the villa, and were tethered in the shade around the houses; and then I had to creep about like a thief by back ways, for no one could tell what mischief they might do to children and houses were I to walk among them. If I came suddenly upon a well where women were drawing water or children bathing, a sudden flight was the certain result; which things occurring day after day, were very unpleasant to a person who does not like to be disliked, and who had never been accustomed to be treated as an ogre.

About the middle of November, finding my health no better, and insects, birds, and shells all very scarce, I determined to return to Mamajam, and pack up my collections before the heavy rains commenced. The wind had already begun to blow from the west, and many signs indicated that the rainy season might set in earlier than usual; and then everything becomes very damp, and it is almost impossible to dry collections properly. My kind friend Mr. Mesman again lent me his pack-horses, and with the assistance of a few men to carry my birds and insects, which I did not like to trust on horses' backs, we got everything home safe. Few can imagine the luxury it was to stretch myself on a sofa, and to take my supper comfortably at table seated in my easy bamboo chair, after having for five weeks taken all my meals uncomfortably on the floor. Such things are trifles in health, but when the body is weakened by disease the habits of a lifetime cannot be so easily set aside.

My house, like all bamboo structures in this country, was a leaning one, the strong westerly winds of the wet season having set all its posts out of the perpendicular to such a degree as to make me think it might someday possibly go over altogether. It is a remarkable thing that the natives of Celebes have not discovered the use of diagonal struts in strengthening buildings. I doubt if there is a native house in the country two years old and at all exposed to the wind, which stands upright; and no wonder, as they merely consist of posts and joists all placed upright or horizontal, and fastened rudely together with rattans. They may be seen in every stage of the process of tumbling down, from the first slight inclination, to such a dangerous slope that it becomes a notice to quit to the occupiers.

The mechanical geniuses of the country have only discovered two ways of remedying the evil. One is, after it has commenced, to tie the house to a post in the ground on the windward side by a rattan or bamboo cable. The other is a preventive, but how they ever found it out and did not discover the true way is a mystery. This plan is, to build the house in the usual way, but instead of having all the principal supports of straight posts, to have two or three of them chosen as crooked as possible. I had often noticed these crooked posts in houses, but imputed it to the scarcity of good, straight timber, until one day I met some men carrying home a post shaped something like a dog's hind leg, and inquired of my native boy what they were going to do with such a piece of wood. "To make a post for a house," said he. "But why don't they get a straight one, there are plenty here?" said I. "Oh," replied he, "they prefer some like that in a house, because then it won't fall," evidently imputing the effect to some occult property of crooked timber. A little consideration and a diagram will, however, show, that the effect imputed to the crooked post may be really produced by it. A true square changes its figure readily into a rhomboid or oblique figure, but when one or two of the uprights are bent or sloping, and placed so as to oppose each other, the effect of a strut is produced, though in a rude and clumsy manner.

Just before I had left Mamajam the people had sown a considerable quantity of maize, which appears above ground in two or three days, and in favourable seasons ripens in less than two months. Owing to a week's premature rains the ground was all flooded when I returned, and the plants just coming into ear were yellow and dead. Not a grain would be obtained by the whole village, but luckily it is only a luxury, not a necessity of life. The rain was the signal for ploughing to begin, in order to sow rice on all the flat lands between us and the town. The plough used is a rude wooden instrument with a very short single handle, a tolerably well-shaped coulter, and the point formed of a piece of hard palm-wood fastened in with wedges. One or two buffaloes draw it at a very slow pace. The seed is sown broadcast, and a rude wooden harrow is used to smooth the surface.

By the beginning of December the regular wet season had set in. Westerly winds and driving rains sometimes continued for days together; the fields for miles around were under water, and the ducks and buffaloes enjoyed themselves amazingly. All along the road to Macassar, ploughing was daily going on in the mud and water, through which the wooden plough easily makes its way, the ploughman holding the plough-handle with one hand while a long bamboo in the other serves to guide the buffaloes. These animals require an immense deal of driving to get them on at all; a continual shower of exclamations is kept up at them, and "Oh! ah! Gee! ugh!" are to be heard in various keys and in an uninterrupted succession all day long. At night we were favoured with a different kind of concert. The dry ground around my house had become a marsh tenanted by frogs, who kept up a most incredible noise from dusk to dawn. They were somewhat musical too, having a deep vibrating note which at times closely resembles the tuning of two or three bass-viols in an orchestra. In Malacca and Borneo I had heard no such sounds as these, which indicates that the frogs, like most of the animals of Celebes, are of species peculiar to it.

My kind friend and landlord, Mr. Mesman, was a good specimen of the Macassar-born Dutchman. He was about thirty-five years of age, had a large family, and lived in a spacious house near the town, situated in the midst of a grove of fruit trees, and surrounded by a perfect labyrinth of offices, stables, and native cottages occupied by his numerous servants, slaves, or dependants. He usually rose before the sun, and after a cup of coffee looked after his servants, horses, and dogs, until seven, when a substantial breakfast of rice and meat was ready in a cool verandah. Putting on a clean white linen suit, he then drove to town in his buggy, where he had an office, with two or three Chinese clerks who looked after his affairs. His business was that of a coffee and opium merchant. He had a coffee estate at Bontyne, and a small prau which traded to the Eastern islands near New Guinea, for mother-of-pearl and tortoiseshell. About one he would return home, have coffee and cake or fried plantain, first changing his dress for a coloured cotton shirt and trousers and bare feet, and then take a siesta with a book. About four, after a cup of tea, he would walk round his premises, and generally stroll down to Mamajam to pay me a visit, and look after his farm.

This consisted of a coffee plantation and an orchard of fruit trees, a dozen horses and a score of cattle, with a small village of Timorese slaves and Macassar servants. One family looked after the cattle and supplied the house with milk, bringing me also a large glassful every morning, one of my greatest luxuries. Others had charge of the horses, which were brought in every afternoon and fed with cut grass. Others had to cut grass for their master's horses at Macassar—not a very easy task in the dry season, when all the country looks like baked mud; or in the rainy season, when miles in every direction are flooded. How they managed it was a mystery to me, but they know grass must be had, and they get it. One lame woman had charge of a flock of ducks. Twice a day she took them out to feed in the marshy places, let them waddle and gobble for an hour or two, and then drove them back and shut them up in a small dark shed to digest their meal, whence they gave forth occasionally a melancholy quack. Every night a watch was set, principally for the sake of the horses—the people of Goa, only two miles off, being notorious thieves, and horses offering the easiest and most valuable spoil. This enabled me to sleep in security, although many people in Macassar thought I was running a great risk, living alone in such a solitary place and with such bad neighbours.

My house was surrounded by a kind of straggling hedge of roses, jessamines, and other flowers, and every morning one of the women gathered a basketful of the blossoms for Mr. Mesman's family. I generally took a couple for my own breakfast table, and the supply never failed during my stay, and I suppose never does. Almost every Sunday Mr. M. made a shooting excursion with his eldest son, a lad of fifteen, and I generally accompanied him; for though the Dutch are Protestants, they do not observe Sunday in the rigid manner practised in England and English colonies. The Governor of the place has his public reception every Sunday evening, when card-playing is the regular amusement.

On December 13th I went on board a prau bound for the Aru Islands, a journey which will be described in the latter part of this work.

On my return, after a seven months' absence, I visited another district to the north of Macassar, which will form the subject of the next CHAPTER.

CHAPTER XVI. CELEBES.

(MACASSAR, JULY TO NOVEMBER, 1857.)

I REACHED Macassar again on the 11th of July, and established myself in my old quarters at Mamajam, to sort, arrange, clean, and pack up my Aru collections. This occupied me a month; and having shipped them off for Singapore, had my guns repaired, and received a new one from England, together with a stock of pins, arsenic, and other collecting requisites. I began to feel eager for work again, and had to consider where I should spend my time until the end of the year; I had left Macassar seven months before, a flooded marsh being ploughed up for the rice-sowing. The rains had continued for five months, yet now all the rice was cut, and dry and dusty stubble covered the country just as when I had first arrived there.

After much inquiry I determined to visit the district of Maros, about thirty miles north of Macassar, where Mr. Jacob Mesman, a brother of my friend, resided, who had kindly offered to find me house-room and give me assistance should I feel inclined to visit him. I accordingly obtained a pass from the Resident, and having hired a boat set off one evening for Maros. My boy Ali was so ill with fever that I was obliged to leave him in the hospital, under the care of my friend the German doctor, and I had to make shift with two new servants utterly ignorant of everything. We coasted along during the night, and at daybreak entered the Maros river, and by three in the afternoon reached the village. I immediately visited the Assistant Resident, and applied for ten men to carry my baggage, and a horse for myself. These were promised to be ready that night, so that I could start as soon as I liked in the morning. After having taken a cup of tea I took my leave, and slept in the boat. Some of the men came at night as promised, but others did not arrive until the next morning. It took some time to divide my baggage fairly among them, as they all wanted to shirk the heavy boxes, and would seize hold of some light article and march off with it, until made to come back and wait until the whole had been fairly apportioned. At length about eight o'clock all was arranged, and we started for our walk to Mr. M.'s farm.

The country was at first a uniform plain of burned-up rice-grounds, but at a few miles' distance precipitous hills appeared, backed by the lofty central range of the peninsula. Towards these our path lay, and after having gone six or eight miles the hills began to advance into the plain right and left of us, and the ground became pierced here and there with blocks and pillars of limestone rock, while a few abrupt conical hills and peaks rose like islands. Passing over an elevated tract forming the shoulder of one of the hills, a picturesque scene lay before us. We looked down into a little valley almost entirely surrounded by mountains, rising abruptly in huge precipices, and forming a succession of knolls and peaks and domes of the most varied and fantastic shapes. In the very centre of the valley was a large bamboo house, while scattered around were a dozen cottages of the same material.

I was kindly received by Mr. Jacob Mesman in an airy saloon detached from the house, and entirely built of bamboo and thatched with grass. After breakfast he took me to his foreman's house, about a hundred yards off, half of which was given up to me until I should decide where to have a cottage built for my own use. I soon found that this spot was too much exposed to the wind and dust, which rendered it very difficult to work with papers or insects. It was also dreadfully hot in the afternoon, and after a few days I got a sharp attack of fever, which determined me to move. I accordingly fixed on a place about a mile off, at the foot of a forest-covered hill, where in a few days Mr. M. built for me a nice little house, consisting of a good-sized enclosed verandah or open room, and a small inner sleeping-room, with a little cookhouse outside. As soon as it was finished I moved into it, and found the change most agreeable.

The forest which surrounded me was open and free from underwood, consisting of large trees, widely scattered with a great quantity of palm-trees (Arenga saccharifera), from which palm wine and sugar are made. There were also great numbers of a wild Jack-fruit tree (Artocarpus), which bore abundance of large reticulated fruit, serving as an excellent vegetable. The ground was as thickly covered with dry leaves as it is in an English wood in November; the little rocky streams were all dry, and scarcely a drop of water or even a damp place was anywhere to be seen. About fifty yards below my house, at the foot of the hill, was a deep hole in a watercourse where good water was to be had, and where I went daily to bathe by having buckets of water taken out and pouring it over my body.

My host Mr. M. enjoyed a thoroughly country life, depending almost entirely on his gun and dogs to supply his table. Wild pigs of large size were very plentiful and he generally got one or two a week, besides deer occasionally, and abundance of jungle-fowl, hornbills, and great fruit pigeons. His buffaloes supplied plenty of milk from which he made his own butter; he grew his own rice and coffee, and had ducks, fowls, and their eggs, in profusion. His palm-trees supplied him all the year round with "sagueir," which takes the place of beer; and the sugar made from them is an excellent sweetmeat. All the fine tropical vegetables and fruits were abundant in their season, and his cigars were made from tobacco of his own raising. He kindly sent me a bamboo of buffalo-milk every morning; it was as thick as cream, and required diluting with water to keep it fluid during the day. It mixes very well with tea and coffee, although it has a slight peculiar flavour, which after a time is not disagreeable. I also got as much sweet "sagueir" as I liked to drink, and Mr. M. always sent me a piece of each pig he killed, which with fowls, eggs, and the birds we shot ourselves, and buffalo beef about once a fortnight, kept my larder sufficiently well supplied.

Every bit of flatland was cleared and used as rice-fields, and on the lower slopes of many of the hills tobacco and vegetables were grown. Most of the slopes are covered with huge blocks of rock, very fatiguing to scramble over, while a number of the hills are so precipitous as to be quite inaccessible. These circumstances, combined with the excessive drought, were very unfavourable for my pursuits. Birds were scarce, and I got but few new to me. Insects were tolerably plentiful, but unequal. Beetles, usually so numerous and interesting, were exceedingly scarce, some of the families being quite absent and others only represented by very minute species. The Flies and Bees, on the other hand, were abundant, and of these I daily obtained new and interesting species. The rare and beautiful Butterflies of Celebes were the chief object of my search, and I found many species altogether new to me, but they were generally so active and shy as to render their capture a matter of great difficulty. Almost the only good place for them was in the dry beds of the streams in the forest, where, at damp places, muddy pools, or even on the dry rocks, all sorts of insects could be found. In these rocky forests dwell some of the finest butterflies in the world. Three species of Ornithoptera, measuring seven or eight inches across the wings, and beautifully marked with spots or masses of satiny yellow on a black ground, wheel through the thickets with a strong sailing flight. About the damp places are swarms of the beautiful blue-banded Papilios, miletus and telephus, the superb golden green P. macedon, and the rare little swallow-tail Papilio rhesus, of all of which, though very active, I succeeded in capturing fine series of specimens.

I have rarely enjoyed myself more than during my residence here. As I sat taking my coffee at six in the morning, rare birds would often be seen on some tree close by, when I would hastily sally out in my slippers, and perhaps secure a prize I had been seeking after for weeks. The great hornbills of Celebes (Buceros cassidix) would often come with loud-flapping wings, and perch upon a lofty tree just in front of me; and the black baboon-monkeys, Cynopithecus nigrescens, often stared down in astonishment at such an intrusion into their domains while at night herds of wild pigs roamed about the house, devouring refuse, and obliging us to put away everything eatable or breakable from our little cooking-house. A few minutes' search on the fallen trees around my house at sunrise and sunset, would often produce me more beetles than I would meet with in a day's collecting, and odd moments could be made valuable which when living in villages or at a distance from the forest are inevitably wasted. Where the sugar-palms were dripping with sap, flies congregated in immense numbers, and it was by spending half an hour at these when I had the time to spare, that I obtained the finest and most remarkable collection of this group of insects that I have ever made.

Then what delightful hours I passed wandering up and down the dry river-courses, full of water-holes and rocks and fallen trees, and overshadowed by magnificent vegetation. I soon got to know every hole and rock and stump, and came up to each with cautious step and bated breath to see what treasures it would produce. At one place I would find a little crowd of the rare butterfly Tachyris zarinda, which would rise up at my approach, and display their vivid orange and cinnabar-red wings, while among them would flutter a few of the fine blue-banded Papilios. Where leafy branches hung over the gully, I might expect to find a grand Ornithoptera at rest and an easy prey. At certain rotten trunks I was sure to get the curious little tiger beetle, Therates flavilabris. In the denser thickets I would capture the small metal-blue butterflies (Amblypodia) sitting on the leaves, as well as some rare and beautiful leaf-beetles of the families Hispidae and Chrysomelidae.

I found that the rotten jack-fruits were very attractive to

many beetles, and used to split them partly open and lay them about in the forest near my house to rot. A morning's search at these often produced me a score of species—Staphylinidae, Nitidulidae, Onthophagi, and minute Carabidae, being the most abundant. Now and then the "sagueir" makers brought me a fine rosechafer (Sternoplus schaumii) which they found licking up the sweet sap. Almost the only new birds I met with for some time were a handsome ground thrush (Pitta celebensis), and a beautiful violet-crowned dove (Ptilonopus celebensis), both very similar to birds I had recently obtained at Aru, but of distinct species.

About the latter part of September a heavy shower of rain fell, admonishing us that we might soon expect wet weather, much to the advantage of the baked-up country. I therefore determined to pay a visit to the falls of the Maros river, situated at the point where it issues from the mountains—a spot often visited by travellers and considered very beautiful. Mr. M. lent me a horse, and I obtained a guide from a neighbouring village; and taking one of my men with me, we started at six in the morning, and after a ride of two hours over the flat rice-fields skirting the mountains which rose in grand precipices on our left, we reached the river about half-way between Maros and the falls, and thence had a good bridle-road to our destination, which we reached in another hour. The hills had closed in around us as we advanced; and when we reached a ruinous shed which had been erected for the accommodation of visitors, we found ourselves in a flat-bottomed valley about a quarter of a mile wide, bounded by precipitous and often overhanging limestone rocks. So far the ground had been cultivated, but it now became covered with bushes and large scattered trees.

As soon as my scanty baggage had arrived and was duly deposited in the shed, I started off alone for the fall, which was about a quarter of a mile further on. The river is here about twenty yards wide, and issues from a chasm between two vertical walls of limestone, over a rounded mass of basaltic rock about forty feet high, forming two curves separated by a slight ledge. The water spreads beautifully over this surface in a thin sheet of foam, which curls and eddies in a succession of concentric cones until it falls into a fine deep pool below. Close to the very edge of the fall a narrow and very rugged path leads to the river above, and thence continues close under the precipice along the water's edge, or sometimes in the water, for a few hundred yards, after which the rocks recede a little, and leave a wooded bank on one side, along which the path is continued, until in about half a mile, a second and smaller fall is reached. Here the river seems to issue from a cavern, the rocks having fallen from above so as to block up the channel and bar further progress. The fall itself can only be reached by a path which ascends behind a huge slice of rock which has partly fallen away from the mountain, leaving a space two or three feet wide, but disclosing a dark chasm descending into the bowels of the mountain, and which, having visited several such, I had no great curiosity to explore.

Crossing the stream a little below the upper fall, the path ascends a steep slope for about five hundred feet, and passing through a gap enters a narrow valley, shut in by walls of rock absolutely perpendicular and of great height. Half a mile further this valley turns abruptly to the right, and becomes a mere rift in the mountain. This extends another half mile, the walls gradually approaching until they are only two feet apart, and the bottom rising steeply to a pass which leads probably into another valley, but which I had no time to explore. Returning to where this rift had begun the main path turns up to the left in a sort of gully, and reaches a summit over which a fine natural arch of rock passes at a height of about fifty feet. Thence was a steep descent through thick jungle with glimpses of precipices and distant rocky mountains, probably leading into the main river valley again. This was a most tempting region to explore, but there were several reasons why I could go no further. I had no guide, and no permission to enter the Bugis territories, and as the rains might at any time set in, I might be prevented from returning by the flooding of the river. I therefore devoted myself during the short time of my visit to obtaining what knowledge I could of the natural productions of the place.

The narrow chasms produced several fine insects quite

new to me, and one new bird, the curious Phlaegenas tristigmata, a large ground pigeon with yellow breast and crown, and purple neck. This rugged path is the highway from Maros to the Bugis country beyond the mountains. During the rainy season it is quite impassable, the river filling its bed and rushing between perpendicular cliffs many hundred feet high. Even at the time of my visit it was most precipitous and fatiguing, yet women and children came over it daily, and men carrying heavy loads of palm sugar (of very little value). It was along the path between the lower and the upper falls, and about the margin of the upper pool, that I found most insects. The large semi-transparent butterfly, Idea tondana, flew lazily along by dozens, and it was here that I at length obtained an insect which I had hoped but hardly expected to meet with—the magnificent Papilio androcles, one of the largest and rarest known swallow-tailed butterflies. During my four days' stay at the falls, I was so fortunate as to obtain six good specimens. As this beautiful creature flies, the long white tails flicker like streamers, and when settled on the beach it carries them raised upwards, as if to preserve them from injury. It is scarce even here, as I did not see more than a dozen specimens in all, and had to follow many of them up and down the river's bank repeatedly before I succeeded in their capture. When the sun shone hottest, about noon, the moist beach of the pool below the upper fall presented a beautiful sight, being dotted with groups of gay butterflies—orange, yellow, white, blue, and green—which on being disturbed rose into the air by hundreds, forming clouds of variegated colours.

Such gorges, chasms, and precipices here abound, as I have nowhere seen in the Archipelago. A sloping surface is scarcely anywhere to be found, huge walls and rugged masses of rock terminating all the mountains and enclosing the valleys. In many parts there are vertical or even overhanging precipices five or six hundred feet high, yet completely clothed with a tapestry of vegetation. Ferns, Pandanaceae, shrubs, creepers, and even forest trees, are mingled in an evergreen network, through the interstices of which appears the white limestone rock or the dark holes and chasms with which it abounds. These precipices are enabled to sustain such an amount of vegetation by their peculiar structure. Their surfaces are very irregular, broken into holes and fissures, with ledges overhanging the mouths of gloomy caverns; but from each projecting part have descended stalactites, often forming a wild gothic tracery over the caves and receding hollows, and affording an admirable support to the roots of the shrubs, trees, and creepers, which luxuriate in the warm pure atmosphere and the gentle moisture which constantly exudes from the rocks. In places where the precipice offers smooth surfaces of solid rock, it remains quite bare, or only stained with lichens, and dotted with clumps of ferns that grow on the small ledges and in the minutest crevices.

Volume I

The reader who is familiar with tropical nature only through the medium of books and botanical gardens will picture to himself in such a spot many other natural beauties. He will think that I have unaccountably forgotten to mention the brilliant flowers, which, in gorgeous masses of crimson, gold or azure, must spangle these verdant precipices, hang over the cascade, and adorn the margin of the mountain stream. But what is the reality? In vain did I gaze over these vast walls of verdure, among the pendant creepers and bushy shrubs, all around the cascade on the river's bank, or in the deep caverns and gloomy fissures—not one single spot of bright colour could be seen, not one single tree or bush or creeper bore a flower sufficiently conspicuous to form an object in the landscape. In every direction the eye rested on green foliage and mottled rock. There was infinite variety in the colour and aspect of the foliage; there was grandeur in the rocky masses and in the exuberant luxuriance of the vegetation; but there was no brilliancy of colour, none of those bright flowers and gorgeous masses of blossom so generally considered to be everywhere present in the tropics. I have here given an accurate sketch of a luxuriant tropical scene as noted down on the spot, and its general characteristics as regards colour have been so often repeated, both in South America and over many thousand miles in the Eastern tropics, that I am driven to conclude that it represents the general aspect of nature at the equatorial (that is, the most tropical) parts of the tropical regions.

How is it then, that the descriptions of travellers generally give a very different idea? and where, it may be asked, are the glorious flowers that we know do exist in the tropics? These questions can be easily answered. The fine tropical flowering-plants cultivated in our hothouses have been culled from the most varied regions, and therefore give a most erroneous idea of their abundance in any one region. Many of them are very rare, others extremely local, while a considerable number inhabit the more arid regions of Africa and India, in which tropical vegetation does not exhibit itself in its usual luxuriance. Fine and varied foliage, rather than gay flowers, is more characteristic of those parts where tropical vegetation attains its highest development, and in such districts each kind of flower seldom lasts in perfection more than a few weeks, or sometimes a few days. In every locality a lengthened residence will show an abundance of magnificent and gaily-blossomed plants, but they have to be sought for, and are rarely at any one time or place so abundant as to form a perceptible feature in the landscape. But it has been the custom of travellers to describe and group together all the fine plants they have met with during a long journey, and thus produce the effect of a gay and flower-painted landscape. They have rarely studied and described individual scenes where vegetation was most luxuriant and beautiful, and fairly stated what effect was produced in them by flowers. I have done so frequently, and the result of these examinations has convinced me that the bright colours of flowers have a much greater influence on the general aspect of nature in temperate than in tropical climates. During twelve years spent amid the grandest tropical vegetation, I have seen nothing comparable to the effect produced on our landscapes by gorse, broom, heather, wild hyacinths, hawthorn, purple orchises, and buttercups.

The geological structure of this part of Celebes is interesting. The limestone mountains, though of great extent, seem to be entirely superficial, resting on a basis of basalt which in some places forms low rounded hills between the more precipitous mountains. In the rocky beds of the streams basalt is almost always found, and it is a step in this rock which forms the cascade already described. From it the limestone precipices rise abruptly; and in ascending the little stairway along the side of the fall, you step two or three times from one rock on to the other—the limestone dry and rough, being worn by the water and rains into sharp ridges and honeycombed holes—the basalt moist, even, and worn smooth and slippery by the passage of bare-footed pedestrians. The solubility of the limestone by rain-water is well seen in the little blocks and peaks which rise thickly through the soil of the alluvial plains as you approach the mountains. They are all skittle-shaped, larger in the middle than at the base, the greatest diameter occurring at the height to which the country is flooded in the wet season, and thence decreasing regularly to the ground. Many of them overhang considerably, and some of the slenderer pillars appear to stand upon a point. When the rock is less solid it becomes curiously honeycombed by the rains of successive winters, and I noticed some masses reduced to a complete network of stone through which light could be seen in every direction.

From these mountains to the sea extends a perfectly flat alluvial plain, with no indication that water would accumulate at a great depth beneath it, yet the authorities at Macassar have spent much money in boring a well a thousand feet deep in hope of getting a supply of water like that obtained by the Artesian wells in the London and Paris basins. It is not to be wondered at that the attempt was unsuccessful.

Returning to my forest hut, I continued my daily search after birds and insects. The weather, however, became dreadfully hot and dry, every drop of water disappearing from the pools and rock-holes, and with it the insects which frequented them. Only one group remained unaffected by the intense drought; the Diptera, or two-winged flies, continued as plentifully as ever, and on these I was almost compelled to concentrate my attention for a week or two, by which means I increased my collection of that Order to about two hundred species. I also continued to obtain a few new birds, among which were two or three kinds of small hawks and falcons, a beautiful brush-tongued paroquet, Trichoglossus ornatus, and a rare black and white crow, Corvus advena.

At length, about the middle of October, after several gloomy days, down came a deluge of rain which continued to fall almost every afternoon, showing that the early part of the wet season had commenced. I hoped now to get a good harvest of insects, and in some respects I was not disappointed. Beetles became much more numerous, and under a thick bed of leaves that had accumulated on some rocks by the side of a forest stream, I found an abundance of Carabidae, a family generally scarce in the tropics. The butterflies, however, disappeared. Two of my servants were attacked with fever, dysentery, and swelled feet, just at the time that the third had left me, and for some days they both lay groaning in the house. When they got a little better I was attacked myself, and as my stores were nearly finished and everything was getting very damp, I was obliged to prepare for my return to Macassar, especially as the strong westerly winds would render the passage in a small open boat disagreeable, if not dangerous.

Since the rains began, numbers of huge millipedes, as thick as one's finger and eight or ten inches long, crawled about everywhere—in the paths, on trees, about the house—and one morning when I got up I even found one in my bed! They were generally of a dull lead colour or of a deep brick red, and were very nasty-looking things to be coming everywhere in one's way, although quite harmless.

Snakes too began to show themselves. I killed two of a very abundant species—big-headed, and of a bright green colour, which lie coiled up on leaves and shrubs and can scarcely be seen until one is close upon them. Brown snakes got into my net while beating among dead leaves for insects, and made me rather cautious about inserting my hand until I knew what kind of game I had captured. The fields and meadows which had been parched and sterile, now became suddenly covered with fine long grass; the river-bed where I had so many times walked over burning rocks, was now a deep and rapid stream; and numbers of herbaceous plants and shrubs were everywhere springing up and bursting into flower. I found plenty of new insects, and if I had had a good, roomy, water-and-wind-proof house, I should perhaps have stayed during the wet season, as I feel sure many things can then be obtained which are to be found at no other time. With my summer hut, however, this was impossible. During the heavy rains a fine drizzly mist penetrated into every part of it, and I began to have the greatest difficulty in keeping my specimens dry.

Early in November I returned to Macassar, and having packed up my collections, started in the Dutch mail steamer for Amboyna and Ternate. Leaving this part of my journey for the present, I will in the next CHAPTER conclude my account of Celebes, by describing the extreme northern part of the island which I visited two years later.

CHAPTER XVII. CELEBES.

(MENADO, JUNE TO SEPTEMBER, 1859.)

IT was after my residence at Timor-Coupang that I visited the northeastern extremity of Celebes, touching Banda, Amboyna, and Ternate on my way. I reached Menado on the 10th of June, 1859, and was very kindly received by Mr. Tower, an Englishman, but a very old resident in Menado, where he carries on a general business. He introduced me to Mr. L. Duivenboden (whose father had been my friend at Ternate), who had much taste for natural history; and to Mr. Neys, a native of Menado, but who was educated at Calcutta, and to whom Dutch, English, and Malay were equally mother-tongues. All these gentlemen showed me the greatest kindness, accompanied me in my earliest walks about the country, and assisted me by every means in their power. I spent a week in the town very pleasantly, making explorations and inquiries after a good collecting station, which I had much difficulty in finding, owing to the wide cultivation of coffee and cacao, which has led to the clearing away of the forests for many miles around the town, and over extensive districts far into the interior.

The little town of Menado is one of the prettiest in the

East. It has the appearance of a large garden containing rows of rustic villas with broad paths between, forming streets generally at right angles with each other. Good roads branch off in several directions towards the interior, with a succession of pretty cottages, neat gardens, and thriving plantations, interspersed with wildernesses of fruit trees. To the west and south the country is mountainous, with groups of fine volcanic peaks 6,000 or 7,000 feet high, forming grand and picturesque backgrounds to the landscape.

The inhabitants of Minahasa (as this part of Celebes is called) differ much from those of all the rest of the island, and in fact from any other people in the Archipelago. They are of a light-brown or yellow tint, often approaching the fairness of a European; of a rather short stature, stout and well-made; of an open and pleasing countenance, more or less disfigured as age increases by projecting check-bones; and with the usual long, straight, jet-black hair of the Malayan races. In some of the inland villages where they may be supposed to be of the purest race, both men and women are remarkably handsome; while nearer the coasts where the purity of their blood has been destroyed by the intermixture of other races, they approach to the ordinary types of the wild inhabitants of the surrounding countries.

In mental and moral characteristics they are also highly peculiar. They are remarkably quiet and gentle in disposition, submissive to the authority of those they consider their superiors, and easily induced to learn and adopt the habits of civilized people. They are clever mechanics, and seem capable of acquiring a considerable amount of intellectual education.

Up to a very recent period these people were thorough savages, and there are persons now living in Menado who remember a state of things identical with that described by the writers of the sixteenth and seventeenth centuries. The inhabitants of the several villages were distinct tribes, each under its own chief, speaking languages unintelligible to each other, and almost always at war. They built their houses elevated upon lofty posts to defend themselves from the attacks of their enemies. They were headhunters like the Dyaks of Borneo, and were said to be sometimes cannibals. When a chief died, his tomb was adorned with two fresh human heads; and if those of enemies could not be obtained, slaves were killed for the occasion. Human skulls were the great ornaments of the chiefs' houses. Strips of bark were their only dress. The country was a pathless wilderness, with small cultivated patches of rice and vegetables, or clumps of fruit-trees, diversifying the otherwise unbroken forest. Their religion was that naturally engendered in the undeveloped human mind by the contemplation of grand natural phenomena and the luxuriance of tropical nature. The burning mountain, the torrent and the lake, were the abode of their deities; and certain trees and birds were supposed to have special influence over men's actions and destiny. They held wild and exciting festivals to propitiate these deities or demons, and believed that men could be changed by them into animals—either during life or after death.

Here we have a picture of true savage life; of small isolated communities at war with all around them, subject to the wants and miseries of such a condition, drawing a precarious existence from the luxuriant soil, and living on, from generation to generation, with no desire for physical amelioration, and no prospect of moral advancement.

Such was their condition down to the year 1822, when the coffee-plant was first introduced, and experiments were made as to its cultivation. It was found to succeed admirably from fifteen hundred feet, up to four thousand feet above the sea. The chiefs of villages were induced to undertake its cultivation. Seed and native instructors were sent from Java; food was supplied to the labourers engaged in clearing and planting; a fixed price was established at which all coffee brought to the government collectors was to be paid for, and the village chiefs who now received the titles of "Majors" were to receive five percent of the produce. After a time, roads were made from the port of Menado up to the plateau, and smaller paths were cleared from village to village; missionaries settled in the more populous districts and opened schools; and Chinese traders penetrated to the interior and supplied clothing and other luxuries in exchange for the money which the sale of the coffee had produced.

At the same time, the country was divided into districts, and the system of "Controlleurs," which had worked so well in Java, was introduced. The "Controlleur" was a European, or a native of European blood, who was the general superintendent of the cultivation of the district, the adviser of the chiefs, the protector of the people, and the means of communication between both and the European Government. His duties obliged him to visit every village in succession once a month, and to send in a report on their condition to the Resident. As disputes between adjacent villages were now settled by appeal to a superior authority, the old and inconvenient semi-fortified houses were disused, and under the direction of the "Controlleurs" most of the houses were rebuilt on a neat and uniform plan. It was this interesting district which I was now about to visit.

Having decided on my route, I started at 8 A.M. on the 22d of June. Mr. Tower drove me the first three miles in his chaise, and Mr. Neys accompanied me on horseback three miles further to the village of Lotta. Here we met the Controlleur of the district of Tondano, who was returning home from one of his monthly tours, and who had agreed to act as my guide and companion on the journey. From Lotta we had an almost continual ascent for six miles, which brought us on to the plateau of Tondano at an elevation of about 2,400 feet. We passed through three villages whose neatness and beauty quite astonished me. The main road, along which all the coffee is brought down from the interior in carts drawn by buffaloes, is always turned aside at the entrance of a village, so as to pass behind it, and thus allow the village street itself to be kept neat and clean. This is bordered by neat hedges often formed entirely of rose-trees, which are perpetually in blossom. There is a broad central path and a border of fine turf, which is kept well swept and neatly cut. The houses are all of wood, raised about six feet on substantial posts neatly painted blue, while the walls are whitewashed. They all have a verandah enclosed with a neat balustrade, and are generally surrounded by orange-trees and flowering shrubs. The surrounding scenery is verdant and picturesque. Coffee plantations of extreme luxuriance, noble palms and tree ferns, wooded hills and volcanic peaks, everywhere meet the eye. I had heard much of the beauty of this country, but the reality far surpassed my expectations.

About one o'clock we reached Tomohón, the chief place of a district, having a native chief now called the "Major," at whose house we were to dine. Here was a fresh surprise for me. The house was large, airy and very substantially built of hard native timber, squared and put together in a most workmanlike manner. It was furnished in European style, with handsome chandelier lamps, and the chairs and tables all well made by native workmen. As soon as we entered, madeira and bitters were offered us. Then two handsome boys neatly dressed in white, and with smoothly brushed jet-black hair, handed us each a basin of water and a clean napkin on a salver. The dinner was excellent. Fowls cooked in various ways; wild pig roasted, stewed and fried; a fricassee of bats, potatoes, rice and other vegetables; all served on good china, with finger glasses and fine napkins, and abundance of good claret and beer, seemed to me rather curious at the table of a native chief on the mountains of Celebes. Our host was dressed in a suit of black with patent-leather shoes, and really looked comfortable and almost gentlemanly in them. He sat at the head of the table and did the honours well, though he did not talk much. Our conversation was entirely in Malay, as that is the official language here, and in fact the mother-tongue and only language of the Controlleur, who is a native-born half-breed. The Major's father who was chief before him, wore, I was informed, a strip of bark as his sole costume, and lived in a rude but raised home on lofty poles, and abundantly decorated with human heads. Of course we were expected, and our dinner was prepared in the best style, but I was assured that the chiefs all take a pride in adopting European customs, and in being able to receive their visitors in a handsome manner.

After dinner and coffee, the Controlleur went on to Tondano, and I strolled about the village waiting for my baggage, which was coming in a bullock-cart, and did not arrive until after midnight. Supper was very similar to dinner, and on retiring I found an elegant little room with a comfortable bed, gauze curtains with blue and red hangings, and every convenience. Next morning at sunrise the thermometer in the verandah stood at 69°, which I was told is about the usual lowest temperature at this place, 2,500 feet above the sea. I had a good breakfast of coffee, eggs, and fresh bread and butter, which I took in the spacious verandah amid the odour of roses, jessamine, and other sweet-scented flowers, which filled the garden in front; and about eight o'clock left Tomohón with a dozen men carrying my baggage.

Our road lay over a mountain ridge about 4,000 feet above the sea, and then descended about 500 feet to the little village of Rurúkan, the highest in the district of Minahasa, and probably in all Celebes. Here I had determined to stay for some time to see whether this elevation would produce any change in the zoology. The village had only been formed about ten years, and was quite as neat as those I had passed through, and much more picturesque. It is placed on a small level spot, from which there is an abrupt wooded descent down to the beautiful lake of Tondano, with volcanic mountains beyond. On one side is a ravine, and beyond it a fine mountainous and wooded country.

Near the village are the coffee plantations. The trees are planted in rows, and are kept topped to about seven feet high. This causes the lateral branches to grow very strong, so that some of the trees become perfect hemispheres, loaded with fruit from top to bottom, and producing from ten to twenty pounds each of cleaned coffee annually. These plantations were all formed by the Government, and are cultivated by the villagers under the direction of their chief. Certain days are appointed for weeding or gathering, and the whole working population are summoned by the sound of a gong. An account is kept of the number of hours' work done by each family, and at the year's end, the produce of the sale is divided among them proportionately. The coffee is taken to Government stores established at central places over the whole country, and is paid for at a low fixed price. Out of this a certain percentage goes to the chiefs and majors, and the remainder is divided among the inhabitants. This system works very well, and I believe is at present far better for the people than free-trade would be. There are also large rice-fields, and in this little village of seventy houses, I was informed that a hundred pounds' worth of rice was sold annually.

I had a small house at the very end of the village, almost hanging over the precipitous slope down to the stream, and with a splendid view from the verandah. The thermometer in the morning often stood at 62° and never rose so high as 80°, so that with the thin clothing used in the tropical plains we were always cool and sometimes positively cold, while the spout of water where I went daily for my bath had quite an icy feel. Although I enjoyed myself very much among these fine mountains and forests, I was somewhat disappointed as to my collections. There was hardly any perceptible difference between the animal life in this temperate region and in the torrid plains below, and what difference did exist was in most respects disadvantageous to me. There seemed to be nothing absolutely peculiar to this elevation. Birds and quadrupeds were less plentiful, but of the same species. In insects there seemed to be more difference. The curious beetles of the family Cleridae, which are found chiefly on bark and rotten wood, were finer than I have seen them elsewhere. The beautiful Longicorns were scarcer than usual, and the few butterflies were all of tropical species. One of these, Papilio blumei, of which I obtained a few specimens only, is among the most magnificent I have ever seen. It is a green and gold swallow-tail, with azure-blue and spoon-shaped tails, and was often seen flying about the village when the sun shone, but in a very shattered condition. The great amount of wet and cloudy weather was a great drawback all the time I was at Rurúkan.

Even in the vegetation there is very little to indicate elevation. The trees are more covered with lichens and mosses, and the ferns and tree-ferns are finer and more luxuriant than I had been accustomed to seeing on the low grounds, both probably attributable to the almost perpetual moisture that here prevails. Abundance of a tasteless raspberry, with blue and yellow compositae, have somewhat of a temperate aspect; and minute ferns and Orchideae, with dwarf Begonias on the rocks, make some approach to a sub-alpine vegetation. The forest, however, is most luxuriant. Noble palms, Pandani, and tree-ferns are abundant in it, while the forest trees are completely festooned with Orchideae, Bromeliae, Araceae, Lycopodiums, and mosses. The ordinary stemless ferns abound; some with gigantic fronds ten or twelve feet long, others barely an inch high; some with entire and massive leaves, others elegantly waving their finely-cut foliage, and adding endless variety and interest to the forest paths. The cocoa-nut palm still produces fruit abundantly, but is said to be deficient in oil. Oranges thrive better than below, producing abundance of delicious fruit; but the shaddock or pumplemous (Citrus decumana) requires the full force of a tropical sun, for it will not thrive even at Tondano a thousand feet lower. On the hilly slopes rice is cultivated largely, and ripens well, although the temperature rarely or never rises to 80°, so that one would think it might be grown even in England in fine summers, especially if the young plants were raised under glass.

The mountains have an unusual quantity of earth and vegetable mould spread over them. Even on the steepest slopes there is everywhere a covering of clays and sands, and generally a good thickness of vegetable soil. It is this which perhaps contributes to the uniform luxuriance of the forest, and delays the appearance of that sub-alpine vegetation which depends almost as much on the abundance of rocky and exposed surfaces as on difference of climate. At a much lower elevation on Mount Ophir in Malacca, Dacrydiums and Rhododendrons with abundance of Nepenthes, ferns, and terrestrial orchids suddenly took the place of the lofty forest; but this was plainly due to the occurrence of an extensive slope of bare, granitic rock at an elevation of less than 3,000 feet. The quantity of vegetable soil, and also of loose sands and clays, resting on steep slopes, hill-tops and the sides of ravines, is a curious and important phenomenon. It may be due in part to constant, slight earthquake shocks facilitating the disintegration of rock; but, would also seem to indicate that the country has been long exposed to gentle atmospheric action, and that its elevation has been exceedingly slow and continuous.

During my stay at Rurúkan, my curiosity was satisfied by experiencing a pretty sharp earthquake-shock. On the evening of June 29th, at a quarter after eight, as I was sitting reading, the house began shaking with a very gentle, but rapidly increasing motion. I sat still enjoying the novel sensation for some seconds; but in less than half a minute it became strong enough to shake me in my chair, and to make the house visibly rock about, and creak and crack as if it would fall to pieces. Then began a cry throughout the village of "Tana goyang! tana goyang!" (Earthquake! earthquake!)

Everybody rushed out of their houses—women screamed and children cried—and I thought it prudent to go out too. On getting up, I found my head giddy and my steps unsteady, and could hardly walk without falling. The shock continued about a minute, during which time I felt as if I had been turned round and round, and was almost seasick. Going into the house again, I found a lamp and a bottle of arrack upset. The tumbler which formed the lamp had been thrown out of the saucer in which it had stood. The shock appeared to be nearly vertical, rapid, vibratory, and jerking. It was sufficient, I have no doubt, to have thrown down brick, chimneys, walls, and church towers; but as the houses here are all low, and strongly framed of timber, it is impossible for them to be much injured, except by a shock that would utterly destroy a European city. The people told me it was ten years since they had had a stronger shock than this, at which time many houses were thrown down and some people killed.

At intervals of ten minutes to half an hour, slight shocks and tremors were felt, sometimes strong enough to send us all out again. There was a strange mixture of the terrible and the ludicrous in our situation. We might at any moment have a much stronger shock, which would bring down the house over us, or—what I feared more—cause a landslip, and send us down into the deep ravine on the very edge of which the village is built; yet I could not help laughing each time we ran out at a slight shock, and then in a few moments ran in again. The sublime and the ridiculous were here literally but a step apart. On the one hand, the most terrible and destructive of natural phenomena was in action around us—the rocks, the mountains, the solid earth were trembling and convulsed, and we were utterly impotent to guard against the danger that might at any moment overwhelm us. On the other hand was the spectacle of a number of men, women, and children running in and out of their houses, on what each time proved a very unnecessary alarm, as each shock ceased just as it became strong enough to frighten us. It seemed really very much like "playing at earthquakes," and made many of the people join me in a hearty laugh, even while reminding each other that it really might be no laughing matter.

At length the evening got very cold, and I became very sleepy, and determined to turn in; leaving orders to my boys, who slept nearer the door, to wake me in case the house was in danger of falling. But I miscalculated my apathy, for I could not sleep much. The shocks continued at intervals of half an hour or an hour all night, just strong enough to wake me thoroughly each time and keep me on the alert, ready to jump up in case of danger. I was therefore very glad when morning came. Most of the inhabitants had not been to bed at all, and some had stayed out of doors all night. For the next two days and nights shocks still continued at short intervals, and several times a day for a week, showing that there was some very extensive disturbance beneath our portion of the earth's crust. How vast the forces at work really are can only be properly appreciated when, after feeling their effects, we look abroad over the wide expanse of hill and valley, plain and mountain, and thus realize in a slight degree the immense mass of matter heaved and shaken. The sensation produced by an earthquake is never to be forgotten. We feel ourselves in the grasp of a power to which the wildest fury of the winds and waves are as nothing; yet the effect is more a thrill of awe than the terror which the more boisterous war of the elements produces. There is a mystery and an uncertainty as to the amount of danger we incur, which gives greater play to the imagination, and to the influences of hope and fear. These remarks apply only to a moderate earthquake. A severe one is the most destructive and the most horrible catastrophe to which human beings can be exposed.

A few days after the earthquake I took a walk to Tondano, a large village of about 7,000 inhabitants, situated at the lower end of the lake of the same name. I dined with the Controlleur, Mr. Bensneider, who had been my guide to Tomohón. He had a fine large house, in which he often received visitors; and his garden was the best for flowers which I had seen in the tropics, although there was no great variety. It was he who introduced the rose hedges which give such a charming appearance to the villages; and to him is chiefly due the general neatness and good order that everywhere prevail. I consulted him about a fresh locality, as I found Rurúkan too much in the clouds, dreadfully damp and gloomy, and with a general stagnation of bird and insect life. He recommended me a village some distance beyond the lake, near which was a large forest, where he thought I should find plenty of birds. As he was going himself in a few days, I decided to accompany him.

After dinner I asked him for a guide to the celebrated waterfall on the outlet stream of the lake. It is situated about a mile and half below the village, where a slight rising ground closes in the basin, and evidently once formed, the shore of the lake. Here the river enters a gorge, very narrow and tortuous, along which it rushes furiously for a short distance and then plunges into a great chasm, forming the head of a large valley. Just above the fall the channel is not more than ten feet wide, and here a few planks are thrown across, whence, half hid by luxuriant vegetation, the mad waters may be seen rushing beneath, and a few feet farther plunge into the abyss. Both sight and sound are grand and impressive. It was here that, four years before my visit, the Governor-General of the Netherland Indies committed suicide, by leaping into the torrent. This at least is the general opinion, as he suffered from a painful disease which was supposed to have made him weary of his life. His body was found next day in the stream below.

Unfortunately, no good view of the fall could now be obtained, owing to the quantity of wood and high grass that lined the margins of the precipices. There are two falls, the lower being the most lofty; and it is possible, by long circuit, to descend into the valley and see them from below. Were the best points of view searched for and rendered accessible, these falls would probably be found to be the finest in the Archipelago. The chasm seems to be of great depth, probably 500 or 600 feet. Unfortunately, I had no time to explore this valley, as I was anxious to devote every fine day to increasing my hitherto scanty collections.

Just opposite my abode in Rurúkan was the schoolhouse. The schoolmaster was a native, educated by the Missionary at Tomohón. School was held every morning for about three hours, and twice a week in the evening there was catechising and preaching. There was also a service on Sunday morning. The children were all taught in Malay, and I often heard them repeating the multiplication-table, up to twenty times twenty, very glibly. They always wound up with singing, and it was very pleasing to hear many of our old psalm-tunes in these remote mountains, sung with Malay words. Singing is one of the real blessings which Missionaries introduce among savage nations, whose native chants are almost always monotonous and melancholy.

On catechising evenings the schoolmaster was a great man, preaching and teaching for three hours at a stretch much in the style of an English ranter. This was pretty cold work for his auditors, however warming to himself; and I am inclined to think that these native teachers, having acquired facility of speaking and an endless supply of religious platitudes to talk about, ride their hobby rather hard, without much consideration for their flock. The Missionaries, however, have much to be proud of in this country. They have assisted the Government in changing a savage into a civilized community in a wonderfully short space of time. Forty years ago the country was a wilderness, the people naked savages, garnishing their rude houses with human heads. Now it is a garden, worthy of its sweet native name of "Minahasa." Good roads and paths traverse it in every direction; some of the finest coffee plantations in the world surround the villages, interspersed with extensive rice-fields more than sufficient for the support of the population.

The people are now the most industrious, peaceable, and civilized in the whole Archipelago. They are the best clothed, the best housed, the best fed, and the best educated; and they have made some progress towards a higher social state. I believe there is no example elsewhere of such striking results being produced in so short a time—results which are entirely due to the system of government now adopted by the Dutch in their Eastern possessions. The system is one which may be called a "paternal despotism." Now we Englishmen do not like despotism—we hate the name and the thing, and we would rather see people ignorant, lazy, and vicious, than use any but moral force to make them wise, industrious, and good. And we are right when we are dealing with men of our own race, and of similar ideas and equal capacities with ourselves. Example and precept, the force of public opinion, and the slow, but sure spread of education, will do everything in time, without engendering any of those bitter feelings, or producing any of that servility, hypocrisy, and dependence, which are the sure results of despotic government. But what should we think of a man who should advocate these principles of perfect freedom in a family or a school? We should say that he was applying a good, general principle to a case in which the conditions rendered it inapplicable—the case in which the governed are in an admitted state of mental inferiority to those who govern them, and are unable to decide what is best for their permanent welfare. Children must be subjected to some degree of authority, and guidance; and if properly managed they will cheerfully submit to it, because they know their own inferiority, and believe their elders are acting solely for their good. They learn many things the use of which they cannot comprehend, and which they would never learn without some moral and social, if not physical, pressure. Habits of order, of industry, of cleanliness, of respect and obedience, are inculcated by similar means. Children would never grow up into well-behaved and well-educated men, if the same absolute freedom of action that is allowed to men were allowed to them. Under the best aspect of education, children are subjected to a mild despotism for the good of themselves and of society; and their confidence in the wisdom and goodness of those who ordain and apply this despotism, neutralizes the bad passions and degrading feelings, which under less favourable conditions are its general results.

Now, there is not merely an analogy—there is in many respects an identity of relation between master and pupil or parent and child on the one hand, and an uncivilized race and its civilized rulers on the other. We know (or think we know) that the education and industry, and the common usages of civilized man, are superior to those of savage life; and, as he becomes acquainted with them, the savage himself admits this. He admires the superior acquirements of the civilized man, and it is with pride that he will adopt such usages as do not interfere too much with his sloth, his passions, or his prejudices. But as the willful child or the idle schoolboy, who was never taught obedience, and never made to do anything which of his own free will he was not inclined to do, would in most cases obtain neither education nor manners; so it is much more unlikely that the savage, with all the confirmed habits of manhood and the traditional prejudices of race, should ever do more than copy a few of the least beneficial customs of civilization, without some stronger stimulus than precept, very imperfectly backed by example.

If we are satisfied that we are right in assuming the government over a savage race, and occupying their country, and if we further consider it our duty to do what we can to improve our rude subjects and raise them up towards our own level, we must not be too much afraid of the cry of "despotism" and "slavery," but must use the authority we possess to induce them to do work which they may not altogether like, but which we know to be an indispensable step in their moral and physical advancement. The Dutch have shown much good policy in the means by which they have done this. They have in most cases upheld and strengthened the authority of the native chiefs, to whom the people have been accustomed to render a voluntary obedience; and by acting on the intelligence and self-interest of these chiefs, have brought about changes in the manners and customs of the people, which would have excited ill-feeling and perhaps revolt, had they been directly enforced by foreigners.

In carrying out such a system, much depends upon the character of the people; and the system which succeeds admirably in one place could only be very partially worked out in another. In Minahasa the natural docility and intelligence of the race have made their progress rapid; and how important this is, is well illustrated by the fact, that in the immediate vicinity of the town of Menado are a tribe called Banteks, of a much less tractable disposition, who have hitherto resisted all efforts of the Dutch Government to induce them to adopt any systematic cultivation. These remain in a ruder condition, but engage themselves willingly as occasional porters and labourers, for which their greater strength and activity well adapt them.

No doubt the system here sketched seems open to serious objection. It is to a certain extent despotic, and interferes with free trade, free labour, and free communication. A native cannot leave his village without a pass, and cannot engage himself to any merchant or captain without a Government permit. The coffee has all to be sold to Government, at less than half the price that the local merchant would give for it, and he consequently cries out loudly against "monopoly" and "oppression." He forgets, however, that the coffee plantations were established by the Government at great outlay of capital and skill; that it gives free education to the people, and that the monopoly is in lieu of taxation. He forgets that the product he wants to purchase and make a profit by, is the creation of the Government, without whom the people would still be savages. He knows very well that free trade would, as its first result, lead to the importation of whole cargoes of arrack, which would be carried over the country and exchanged for coffee. That drunkenness and poverty would spread over the land; that the public coffee plantations would not be kept up; that the quality and quantity of the coffee would soon deteriorate; that traders and merchants would get rich, but that the people would relapse into poverty and barbarism. That such is invariably the result of free trade with any savage tribes who possess a valuable product, native or cultivated, is well known to those who have visited such people; but we might even anticipate from general principles that evil results would happen.

If there is one thing rather than another to which the grand law of continuity or development will apply, it is to human progress. There are certain stages through which society must pass in its onward march from barbarism to civilization. Now one of these stages has always been some form or other of despotism, such as feudalism or servitude, or a despotic paternal government; and we have every reason to believe that it is not possible for humanity to leap over this transition epoch, and pass at once from pure savagery to free civilization. The Dutch system attempts to supply this missing link, and to bring the people on by gradual steps to that higher civilization, which we (the English) try to force upon them at once. Our system has always failed. We demoralize and we extirpate, but we never really civilize. Whether the Dutch system can permanently succeed is but doubtful, since it may not be possible to compress the work of ten centuries into one; but at all events it takes nature as a guide, and is therefore, more deserving of success, and more likely to succeed, than ours.

There is one point connected with this question which I think the Missionaries might take up with great physical and moral results. In this beautiful and healthy country, and with abundance of food and necessaries, the population does not increase as it ought to do. I can only impute this to one cause. Infant mortality, produced by neglect while the mothers are working in the plantations, and by general ignorance of the conditions of health in infants. Women all work, as they have always been accustomed to do. It is no hardship to them, but I believe is often a pleasure and relaxation. They either take their infants with them, in which case they leave them in some shady spot on the ground, going at intervals to give them nourishment, or they leave them at home in the care of other children too young to work. Under neither of these circumstances can infants be properly attended to, and great mortality is the result, keeping the increase of population far below the rate which the general prosperity of the country and the universality of marriage would lead us to expect. This is a matter in which the Government is directly interested, since it is by the increase of the population alone that there can be any large and permanent increase in the production of coffee. The Missionaries should take up the question because, by inducing married women to confine themselves to domestic duties, they will decidedly promote a higher civilization, and directly increase the health and happiness of the whole community. The people are so docile and so willing to adopt the manners and customs of Europeans, that the change might be easily effected by merely showing them that it was a question of morality and civilization, and an essential step in their progress towards an equality with their white rulers.

After a fortnight's stay at Rurúkan, I left that pretty and interesting village in search of a locality and climate more productive of birds and insects. I passed the evening with the Controlleur of Tondano, and the next morning at nine, left in a small boat for the head of the lake, a distance of about ten miles. The lower end of the lake is bordered by swamps and marshes of considerable extent, but a little further on, the hills come down to the water's edge and give it very much the appearance of a greet river, the width being about two miles. At the upper end is the village of Kakas, where I dined with the head man in a good house like those I have already described; and then went on to Langówan, four miles distant over a level plain. This was the place where I had been recommended to stay, and I accordingly unpacked my baggage and made myself comfortable in the large house devoted to visitors. I obtained a man to shoot for me, and another to accompany me the next day to the forest, where I was in hopes of finding a good collecting ground.

In the morning after breakfast I started off, but found I had four miles to walk over a wearisome straight road through coffee plantations before I could get to the forest, and as soon as I did so, it came on to rain heavily and did not cease until night. This distance to walk every day was too far for any profitable work, especially when the weather was so uncertain. I therefore decided at once that I must go further on, until I found someplace close to or in a forest country. In the afternoon my friend Mr. Bensneider arrived, together with the Controlleur of the next district, called Belang, from whom I learned that six miles further on there was a village called Panghu, which had been recently formed and had a good deal of forest close to it; and he promised me the use of a small house if I liked to go there.

The next morning I went to see the hot-springs and mud volcanoes, for which this place is celebrated. A picturesque path among plantations and ravines brought us to a beautiful circular basin about forty feet in diameter, bordered by a calcareous ledge, so uniform and truly curved, that it looked like a work of art. It was filled with clear water very near the boiling point, and emitted clouds of steam with a strong sulphureous odour. It overflows at one point and forms a little stream of hot water, which at a hundred yards' distance is still too hot to hold the hand in. A little further on, in a piece of rough wood, were two other springs not so regular in outline, but appearing to be much hotter, as they were in a continual state of active ebullition. At intervals of a few minutes, a great escape of steam or gas took place, throwing up a column of water three or four feet high.

We then went to the mud-springs, which are about a mile off, and are still more curious. On a sloping tract of ground in a slight hollow is a small lake of liquid mud, with patches of blue, red, or white, and in many places boiling and bubbling most furiously. All around on the indurated clay are small wells and craters full of boiling mud. These seem to be forming continually, a small hole appearing first, which emits jets of steam and boiling mud, which upon hardening, forms a little cone with a crater in the middle. The ground for some distance is very unsafe, as it is evidently liquid at a small depth, and bends with pressure like thin ice. At one of the smaller, marginal jets which I managed to approach, I held my hand to see if it was really as hot as it looked, when a little drop of mud that spurted on to my finger scalded like boiling water.

A short distance off, there was a flat bare surface of rock as smooth and hot as an oven floor, which was evidently an old mud-pool, dried up and hardened. For hundreds of yards around where there were banks of reddish and white clay used for whitewash, it was still so hot close to the surface that the hand could hardly bear to be held in cracks a few inches deep, and from which arose a strong sulphureous vapour. I was informed that some years back a French gentleman who visited these springs ventured too near the liquid mud, when the crust gave way and he was engulfed in the horrible caldron.

This evidence of intense heat so near the surface over a large tract of country was very impressive, and I could hardly divest myself of the notion that some terrible catastrophe might at any moment devastate the country. Yet it is probable that all these apertures are really safety-valves, and that the inequalities of the resistance of various parts of the earth's crust will always prevent such an accumulation of force as would be required to upheave and overwhelm any extensive area. About seven miles west of this is a volcano which was in eruption about thirty years before my visit, presenting a magnificent appearance and covering the surrounding country with showers of ashes. The plains around the lake formed by the intermingling and decomposition of volcanic products are of amazing fertility, and with a little management in the rotation of crops might be kept in continual cultivation. Rice is now grown on them for three or four years in succession, when they are left fallow for the same period, after which rice or maize can be again grown. Good rice produces thirty-fold, and coffee trees continue bearing abundantly for ten or fifteen years, without any manure and with scarcely any cultivation.

I was delayed a day by incessant rain, and then proceeded to Panghu, which I reached just before the daily rain began at 11 A.M. After leaving the summit level of the lake basin, the road is carried along the slope of a fine forest ravine. The descent is a long one, so that I estimated the village to be not more than 1,500 feet above the sea, yet I found the morning temperature often 69°, the same as at Tondano at least 600 or 700 feet higher. I was pleased with the appearance of the place, which had a good deal of forest and wild country around it; and found prepared for me a little house consisting only of a verandah and a back room. This was only intended for visitors to rest in, or to pass a night, but it suited me very well. I was so unfortunate, however, as to lose both my hunters just at this time. One had been left at Tondano with fever and diarrhoea, and the other was attacked at Langówan with inflammation of the chest, and as his case looked rather bad I had him sent back to Menado. The people here were all so busy with their rice-harvest, which was important for them to finish owing to the early rains, that I could get no one to shoot for me.

During the three weeks that I stayed at Panghu it rained nearly every day, either in the afternoon only, or all day long; but there were generally a few hours' sunshine in the morning, and I took advantage of these to explore the roads and paths, the rocks and ravines, in search of insects. These were not very abundant, yet I saw enough to convince me that the locality was a good one, had I been there at the beginning instead of at the end of the dry season. The natives brought me daily a few insects obtained at the Sagueir palms, including some fine Cetonias and stag-beetles. Two little boys were very expert with the blowpipe, and brought me a good many small birds, which they shot with pellets of clay. Among these was a pretty little flower-pecker of a new species (Prionochilus aureolimbatus), and several of the loveliest honeysuckers I had yet seen. My general collection of birds was, however, almost at a standstill; for though I at length obtained a man to shoot for me, he was not good for much, and seldom brought me more than one bird a day. The best thing he shot was the large and rare fruit-pigeon peculiar to Northern Celebes (Carpophaga forsteni), which I had long been seeking.

I was myself very successful in one beautiful group of insects, the tiger-beetles, which seem more abundant and varied here than anywhere else in the Archipelago. I first met with them on a cutting in the road, where a hard clayey bank was partially overgrown with mosses and small ferns. Here, I found running about, a small olive-green species which never took flight; and more rarely, a fine purplish black wingless insect, which was always found motionless in crevices, and was therefore, probably nocturnal. It appeared to me to form a new genus. About the roads in the forest, I found the large and handsome Cicindela heros, which I had before obtained sparingly at Macassar; but it was in the mountain torrent of the ravine itself that I got my finest things. On dead trunks overhanging the water and on the banks and foliage, I obtained three very pretty species of Cicindela, quite distinct in size, form, and colour, but having an almost identical pattern of pale spots. I also found a single specimen of a most curious species with very long antennae. But my finest discovery here was the Cicindela gloriosa, which I found on mossy stones just rising above the water. After obtaining my first specimen of this elegant insect, I used to walk up the stream, watching carefully every moss-covered rock and stone. It was rather shy, and would often lead me on a long chase from stone to stone, becoming invisible every time it settled on the damp moss, owing to its rich velvety green colour. On some days I could only catch a few glimpses of it; on others I got a single specimen; and on a few occasions two, but never without a more or less active pursuit. This and several other species I never saw but in this one ravine.

Among the people here I saw specimens of several types, which, with the peculiarities of the languages, gives me some notion of their probable origin. A striking illustration of the low state of civilization of these people, until quite recently, is to be found in the great diversity of their languages. Villages three or four miles apart have separate dialects, and each group of three or four such villages has a distinct language quite unintelligible to all the rest; so that, until the recent introduction of Malay by the Missionaries, there must have been a bar to all free communication. These languages offer many peculiarities. They contain a Celebes-Malay element and a Papuan element, along with some radical peculiarities found also in the languages of the Siau and Sanguir islands further north, and therefore, probably derived from the

Philippine Islands. Physical characteristics correspond. There are some of the less civilized tribes which have semi-Papuan features and hair, while in some villages the true Celebes or Bugis physiognomy prevails. The plateau of Tondano is chiefly inhabited by people nearly as white as the Chinese, and with very pleasing semi-European features. The people of Siau and Sanguir much resemble these, and I believe them to be perhaps immigrants from some of the islands of North Polynesia. The Papuan type will represent the remnant of the aborigines, while those of the Bugis character show the extension northward of the superior Malay races.

As I was wasting valuable time at Panghu, owing to the bad weather and the illness of my hunters, I returned to Menado after a stay of three weeks. Here I had a little touch of fever, and what with drying and packing all of my collections and getting fresh servants, it was a fortnight before I was again ready to start. I now went eastward over an undulating country skirting the great volcano of Klabat, to a village called Lempias, situated close to the extensive forest that covers the lower slopes of that mountain. My baggage was carried from village to village by relays of men; and as each change involved some delay, I did not reach my destination (a distance of eighteen miles) until sunset. I was wet through, and had to wait for an hour in an uncomfortable state until the first installment of my baggage arrived, which luckily contained my clothes, while the rest did not come in until midnight.

This being the district inhabited by that singular annual the Babirusa (Hog-deer), I inquired about skulls and soon obtained several in tolerable condition, as well as a fine one of the rare and curious "Sapi-utan" (Anoa depressicornis). Of this animal I had seen two living specimens at Menado, and was surprised at their great resemblance to small cattle, or still more to the Eland of South Africa. Their Malay name signifies "forest ox," and they differ from very small highbred oxen principally by the low-hanging dewlap, and straight, pointed horns which slope back over the neck. I did not find the forest here so rich in insects as I had expected, and my hunters got me very few birds, but what they did obtain were very interesting. Among these were the rare forest Kingfisher (Cittura cyanotis), a small new species of Megapodius, and one specimen of the large and interesting Maleo (Megacephalon rubripes), to obtain which was one of my chief reasons for visiting this district. Getting no more, however, after ten days' search, I removed to Licoupang, at the extremity of the peninsula, a place celebrated for these birds, as well as for the Babirusa and Sapi-utan. I found here Mr. Goldmann, the eldest son of the Governor of the Moluccas, who was superintending the establishment of some Government salt-works. This was a better locality, and I obtained some fine butterflies and very good birds, among which was one more specimen of the rare ground dove

(Phlegaenas tristigmata), which I had first obtained near the Maros waterfall in South Celebes.

Hearing what I was particularly in search of, Mr. Goldmann kindly offered to make a hunting-party to the place where the "Maleos" are most abundant, a remote and uninhabited sea-beach about twenty miles distant. The climate here was quite different from that on the mountains; not a drop of rain having fallen for four months; so I made arrangements to stay on the beach a week, in order to secure a good number of specimens. We went partly by boat and partly through the forest, accompanied by the Major or head-man of Licoupang, with a dozen natives and about twenty dogs. On the way they caught a young Sapi-utan and five wild pigs. Of the former I preserved the head. This animal is entirely confined to the remote mountain forests of Celebes and one or two adjacent islands which form part of the same group. In the adults the head is black, with a white mark over each eye, one on each cheek and another on the throat. The horns are very smooth and sharp when young, but become thicker and ridged at the bottom with age. Most naturalists consider this curious animal to be a small ox, but from the character of the horns, the fine coat of hair and the descending dewlap, it seemed closely to approach the antelopes.

Arrived at our destination, we built a but and prepared for a stay of some days—I to shoot and skin "Maleos", and

Mr. Goldmann and the Major to hunt wild pigs, Babirusa, and Sapi-utan. The place is situated in the large bay between the islands of Limbe and Banca, and consists of steep beach more than a mile in length, of deep loose and coarse black volcanic sand (or rather gravel), very fatiguing to walk over. It is bounded at each extremity by a small river with hilly ground beyond, while the forest behind the beach itself is tolerably level and its growth stunted. We probably have here an ancient lava stream from the Klabat volcano, which has flowed down a valley into the sea, and the decomposition of which has formed the loose black sand. In confirmation of this view, it may be mentioned that the beaches beyond the small rivers in both directions are of white sand.

It is in this loose, hot, black sand that those singular birds, the "Maleos" deposit their eggs. In the months of August and September, when there is little or no rain, they come down in pairs from the interior to this or to one or two other favourite spots, and scratch holes three or four feet deep, just above high-water mark, where the female deposits a single large egg, which she covers over with about a foot of sand—and then returns to the forest. At the end of ten or twelve days she comes again to the same spot to lay another egg, and each female bird is supposed to lay six or eight eggs during the season. The male assists the female in making the hole, coming down and returning with her. The appearance of the bird when walking on the beach is very handsome. The glossy black and rosy white of the plumage, the helmeted head and elevated tail, like that of the common fowl, give a striking character, which their stately and somewhat sedate walk renders still more remarkable. There is hardly any difference between the sexes, except that the casque or bonnet at the back of the head and the tubercles at the nostrils are a little larger, and the beautiful rosy salmon colour a little deeper in the male bird; but the difference is so slight that it is not always possible to tell a male from a female without dissection. They run quickly, but when shot at or suddenly disturbed, take wing with a heavy noisy flight to some neighbouring tree, where they settle on a low branch; and, they probably roost at night in a similar situation. Many birds lay in the same hole, for a dozen eggs are often found together; and these are so large that it is not possible for the body of the bird to contain more than one fully-developed egg at the same time. In all the female birds which I shot, none of the eggs besides the one large one exceeded the size of peas, and there were only eight or nine of these, which is probably the extreme number a bird can lay in one season.

 Every year the natives come for fifty miles round to obtain these eggs, which are esteemed as a great delicacy, and when quite fresh, are indeed delicious. They are richer than hens' eggs and of a finer favour, and each one completely fills an ordinary teacup, and forms with bread or rice a very good meal. The colour of the shell is a pale brick red, or very rarely pure white. They are elongate and very slightly smaller at one end, from four to four and a half inches long by two and a quarter or two and a half wide.

After the eggs are deposited in the sand, they are no further cared for by the mother. The young birds, upon breaking the shell, work their way up through the sand and run off at once to the forest; and I was assured by Mr. Duivenboden of Ternate, that they can fly the very day they are hatched. He had taken some eggs on board his schooner which hatched during the night, and in the morning the little birds flew readily across the cabin. Considering the great distances the birds come to deposit the eggs in a proper situation (often ten or fifteen miles) it seems extraordinary that they should take no further care of them. It is, however, quite certain that they neither do nor can watch them. The eggs being deposited by a number of hens in succession in the same hole, would render it impossible for each to distinguish its own; and the food necessary for such large birds (consisting entirely of fallen fruits) can only be obtained by roaming over an extensive district, so that if the numbers of birds which come down to this single beach in the breeding season, amounting to many hundreds, were obliged to remain in the vicinity, many would perish of hunger.

In the structure of the feet of this bird, we may detect a cause for its departing from the habits of its nearest allies, the Megapodii and Talegalli, which heap up earth, leaves, stones, and sticks into a huge mound, in which they bury their eggs. The feet of the Maleo are not nearly so large or strong in proportion as in these birds, while its claws are short and straight instead of being long and much curved. The toes are, however, strongly webbed at the base, forming a broad powerful foot, which, with the rather long leg, is well adapted to scratch away the loose sand (which flies up in a perfect shower when the birds are at work), but which could not without much labour accumulate the heaps of miscellaneous rubbish, which the large grasping feet of the Megapodius bring together with ease.

We may also, I think, see in the peculiar organization of the entire family of the Megapodidae or Brush Turkeys, a reason why they depart so widely from the usual habits of the Class of birds. Each egg being so large as entirely to fill up the abdominal cavity and with difficulty pass the walls of the pelvis, a considerable interval is required before the successive eggs can be matured (the natives say about thirteen days). Each bird lays six or eight eggs or even more each season, so that between the first and last there may be an interval of two or three months. Now, if these eggs were hatched in the ordinary way, either the parents must keep sitting continually for this long period, or if they only began to sit after the last egg was deposited, the first would be exposed to injury by the climate, or to destruction by the large lizards, snakes, or other animals which abound in the district; because such large birds must roam about a good deal in search of food. Here then we seem to have a case in which the habits of a bird may be directly traced to its exceptional organization; for it will hardly be maintained that this abnormal structure and peculiar food were given to the Megapodidae in order that they might not exhibit that parental affection, or possess those domestic instincts so general in the Class of birds, and which so much excite our admiration.

It has generally been the custom of writers on Natural History to take the habits and instincts of animals as fixed points, and to consider their structure and organization, as specially adapted, to be in accordance with these. This assumption is however an arbitrary one, and has the bad effect of stifling inquiry into the nature and causes of "instincts and habits," treating them as directly due to a "first cause," and therefore, incomprehensible to us. I believe that a careful consideration of the structure of a species, and of the peculiar physical and organic conditions by which it is surrounded, or has been surrounded in past ages, will often, as in this case, throw much light on the origin of its habits and instincts. These again, combined with changes in external conditions, react upon structure, and by means of "variation" and "natural selection", both are kept in harmony.

My friends remained three days, and got plenty of wild pigs and two Anóas, but the latter were much injured by the dogs, and I could only preserve the heads. A grand hunt which we attempted on the third day failed, owing to bad management in driving in the game, and we waited for five hours perched on platforms in trees without getting a shot, although we had been assured that pigs, Babirusas, and Anóas would rush past us in dozens. I myself, with two men, stayed three days longer to get more specimens of the Maleos, and succeeded in preserving twenty-six very fine ones—the flesh and eggs of which supplied us with abundance of good food.

The Major sent a boat, as he had promised, to take home my baggage, while I walked through the forest with my two boys and a guide, about fourteen miles. For the first half of the distance there was no path, and we had often to cut our way through tangled rattans or thickets of bamboo. In some of our turnings to find the most practicable route, I expressed my fear that we were losing our way, as the sun being vertical, I could see no possible clue to the right direction. My conductors, however, laughed at the idea, which they seemed to consider quite ludicrous; and sure enough, about half way, we suddenly encountered a little hut where people from Licoupang came to hunt and smoke wild pigs. My guide told me he had never before traversed the forest between these two points; and this is what is considered by some travellers as one of the savage "instincts," whereas it is merely the result of wide general knowledge. The man knew the topography of the whole district; the slope of the land, the direction of the streams, the belts of bamboo or rattan, and many other indications of locality and direction; and he was thus enabled to hit straight upon the hut, in the vicinity of which he had often hunted. In a forest of which he knew nothing, he would be quite as much at a loss as a European. Thus it is, I am convinced, with all the wonderful accounts of Indians finding their way through trackless forests to definite points; they may never have passed straight between the two particular points before, but they are well acquainted with the vicinity of both, and have such a general knowledge of the whole country, its water system, its soil and its vegetation, that as they approach the point they are to reach, many easily-recognised indications enable them to hit upon it with certainty.

The chief feature of this forest was the abundance of rattan palms hanging from the trees, and turning and twisting about on the ground, often in inextricable confusion. One wonders at first how they can get into such queer shapes; but it is evidently caused by the decay and fall of the trees up which they have first climbed, after which they grow along the ground until they meet with another trunk up which to ascend. A tangled mass of twisted living rattan, is therefore, a sign that at some former period a large tree has fallen there, though there may be not the slightest vestige of it left. The rattan seems to have unlimited powers of growth, and a single plant may mount up several trees in succession, and thus reach the enormous length they are said sometimes to attain. They much improve the appearance of a forest as seen from the coast; for they vary the otherwise monotonous tree-tops with feathery crowns of leaves rising clear above them, and each terminated by an erect leafy spike like a lightning-conductor.

The other most interesting object in the forest was a beautiful palm, whose perfectly smooth and cylindrical stem rises erect to more than a hundred feet high, with a thickness of only eight or ten inches; while the fan-shaped leaves which compose its crown, are almost complete circles of six or eight feet diameter, borne aloft on long and slender petioles, and beautifully toothed round the edge by the extremities of the leaflets, which are separated only for a few inches from the circumference. It is probably the Livistona rotundifolia of botanists, and is the most complete and beautiful fan-leaf I have ever seen, serving admirably for folding into water-buckets and impromptu baskets, as well as for thatching and other purposes.

A few days afterwards I returned to Menado on horse-back, sending my baggage around by sea; and had just time to pack up all my collections to go by the next mail steamer to Amboyna. I will now devote a few pages to an account

Volume I

of the chief peculiarities of the Zoology of Celebes, and its relation to that of the surrounding countries.

CHAPTER XVIII. NATURAL HISTORY OF CELEBES.

THE position of Celebes is the most central in the Archipelago. Immediately to the north are the Philippine islands; on the west is Borneo; on the east are the Molucca islands; and on the south is the Timor group—and it is on all sides so connected with these islands by its own satellites, by small islets, and by coral reefs, that neither by inspection on the map nor by actual observation around its coast, is it possible to determine accurately which should be grouped with it, and which with the surrounding districts. Such being the case, we should naturally expect to find that the productions of this central island in some degree represented the richness and variety of the whole Archipelago, while we should not expect much individuality in a country, so situated, that it would seem as if it were pre-eminently fitted to receive stragglers and immigrants from all around.

As so often happens in nature, however, the fact turns out to be just the reverse of what we should have expected; and an examination of its animal productions shows Celebes to be at once the poorest in the number of its species, and the most isolated in the character of its productions, of all the great islands in the Archipelago. With its attendant islets it spreads over an extent of sea hardly inferior in length and breadth to that occupied by Borneo, while its actual land area is nearly double that of Java; yet its Mammalia and terrestrial birds number scarcely more than half the species found in the last-named island. Its position is such that it could receive immigrants from every side more readily than Java, yet in proportion to the species which inhabit it, far fewer seem derived from other islands, while far more are altogether peculiar to it; and a considerable number of its animal forms are so remarkable, as to find no close allies in any other part of the world. I now propose to examine the best known groups of Celebesian animals in some detail, to study their relations to those of other islands, and to call attention to the many points of interest which they suggest.

We know far more of the birds of Celebes than we do of any other group of animals. No less than 191 species have been discovered, and though no doubt, many more wading and swimming birds have to be added; yet the list of land birds, 144 in number, and which for our present purpose are much the most important, must be very nearly complete. I myself assiduously collected birds in Celebes for nearly ten months, and my assistant, Mr. Allen, spent two months in the Sula islands. The Dutch naturalist Forsten spent two years in Northern Celebes (twenty years before my visit), and collections of birds had also been sent to Holland from Macassar. The French ship of discovery, L'Astrolabe, also touched at Menado and procured collections. Since my return home, the Dutch naturalists Rosenberg and Bernstein have made extensive collections both in North Celebes and in the Sula islands; yet all their researches combined have only added eight species of land birds to those forming part of my own collection—a fact which renders it almost certain that there are very few more to discover.

Besides Salayer and Boutong on the south, with Peling and Bungay on the east, the three islands of the Sula (or Zula) Archipelago also belong zoologically to Celebes, although their position is such that it would seem more natural to group them with the Moluccas. About 48 land birds are now known from the Sula group, and if we reject from these, five species which have a wide range over the Archipelago, the remainder are much more characteristic of Celebes than of the Moluccas. Thirty-one species are identical with those of the former island, and four are representatives of Celebes forms, while only eleven are Moluccan species, and two more representatives.

But although the Sula islands belong to Celebes, they are so close to Bouru and the southern islands of the Gilolo group, that several purely Moluccan forms have migrated there, which are quite unknown to the island of Celebes itself; the whole thirteen Moluccan species being in this category, thus adding to the productions of Celebes a foreign element which does not really belong to it. In studying the peculiarities of the Celebesian fauna, it will therefore be well to consider only the productions of the main island.

The number of land birds in the island of Celebes is 128, and from these we may, as before, strike out a small number of species which roam over the whole Archipelago (often from India to the Pacific), and which therefore only serve to disguise the peculiarities of individual islands. These are 20 in number, and leave 108 species which we may consider as more especially characteristic of the island. On accurately comparing these with the birds of all the surrounding countries, we find that only nine extend into the islands westward, and nineteen into the islands eastward, while no less than 80 are entirely confined to the Celebesian fauna—a degree of individuality which, considering the situation of the island, is hardly to be equalled in any other part of the world. If we still more closely examine these 80 species, we shall be struck by the many peculiarities of structure they present, and by the curious affinities with distant parts of the world which many of them seem to indicate. These points are of so much interest and importance that it will be necessary to pass in review all those species which are peculiar to the island, and to call attention to whatever is most worthy of remark.

Six species of the Hawk tribe are peculiar to Celebes; three of these are very distinct from allied birds which range over all India to Java and Borneo, and which thus seem to be suddenly changed on entering Celebes. Another (Accipiter trinotatus) is a beautiful hawk, with elegant rows of large round white spots on the tail, rendering it very conspicuous and quite different from any other known bird of the family. Three owls are also peculiar; and one, a barn owl (Strix rosenbergii), is very much larger and stronger than its ally Strix javanica, which ranges from India through all the islands as far as Lombock.

Of the ten Parrots found in Celebes, eight are peculiar. Among them are two species of the singular racquet-tailed parrots forming the genus Prioniturus, and which are characterised by possessing two long spoon-shaped feathers in the tail. Two allied species are found in the adjacent island of Mindanao, one of the Philippines, and this form of tail is found in no other parrots in the whole world. A small species of Lorikeet (Trichoglossus flavoviridis) seems to have its nearest ally in Australia.

The three Woodpeckers which inhabit the island are all peculiar, and are allied to species found in Java and Borneo, although very different from them all.

Among the three peculiar Cuckoos, two are very remarkable. Phoenicophaus callirhynchus is the largest and handsomest species of its genus, and is distinguished by the three colours of its beak, bright yellow, red, and black. Eudynamis melanorynchus differs from all its allies in having a jet-black bill, whereas the other species of the genus always have it green, yellow, or reddish.

The Celebes Roller (Coracias temmincki) is an interesting

example of one species of a genus being cut off from the rest. There are species of Coracias in Europe, Asia, and Africa, but none in the Malay peninsula, Sumatra, Java, or Borneo. The present species seems therefore quite out of place; and what is still more curious is the fact that it is not at all like any of the Asiatic species, but seems more to resemble those of Africa.

In the next family, the Bee-eaters, is another equally isolated bird, Meropogon forsteni, which combines the characters of African and Indian Bee-eaters, and whose only near ally, Meropogon breweri, was discovered by M. Du Chaillu in West Africa!

The two Celebes Hornbills have no close allies in those which abound in the surrounding countries. The only Thrush, Geocichla erythronota, is most nearly allied to a species peculiar to Timor. Two of the Flycatchers are closely allied to Indian species, which are not found in the Malay islands. Two genera somewhat allied to the Magpies (Streptocitta and Charitornis), but whose affinities are so doubtful that Professor Schlegel places them among the Starlings, are entirely confined to Celebes. They are beautiful long-tailed birds, with black and white plumage, and with the feathers of the head somewhat rigid and scale-like.

Doubtfully allied to the Starlings are two other very isolated and beautiful birds. One, Enodes erythrophrys, has ashy and yellow plumage, but is ornamented with broad stripes of orange-red above the eyes. The other, Basilornis celebensis, is a blue-black bird with a white patch on each side of the breast, and the head ornamented with a beautiful compressed scaly crest of feathers, resembling in form that of the well-known Cock-of-the-rock of South America. The only ally to this bird is found in Ceram, and has the feathers of the crest elongated upwards into quite a different form.

A still more curious bird is the Scissirostrum pagei, which although it is at present classed in the Starling family, differs from all other species in the form of the bill and nostrils, and seems most nearly allied in its general structure to the Ox-peckers (Buphaga) of tropical Africa, next to which the celebrated ornithologist Prince Bonaparte finally placed it. It is almost entirely of a slatey colour, with yellow bill and feet, but the feathers of the rump and upper tail-coverts each terminate in a rigid, glossy pencil or tuft of a vivid crimson. These pretty little birds take the place of the metallic-green starlings of the genus Calornis, which are found in most other islands of the Archipelago, but which are absent from Celebes. They go in flocks, feeding upon grain and fruits, often frequenting dead trees, in holes of which they build their nests; and they cling to the trunks as easily as woodpeckers or creepers.

Out of eighteen Pigeons found in Celebes, eleven are peculiar to it. Two of them, Ptilonopus gularis and Turacaena menadensis, have their nearest allies in Timor. Two others,

Carpophaga forsteni and Phlaegenas tristigmata, most resemble Philippine island species; and Carpophaga radiata belongs to a New Guinea group. Lastly, in the Gallinaceous tribe, the curious helmeted Maleo (Megacephalon rubripes) is quite isolated, having its nearest (but still distant) allies in the Brush-turkeys of Australia and New Guinea.

Judging, therefore, by the opinions of the eminent naturalists who have described and classified its birds, we find that many of the species have no near allies whatsoever in the countries which surround Celebes, but are either quite isolated, or indicate relations with such distant regions as New Guinea, Australia, India, or Africa. Other cases of similar remote affinities between the productions of distant countries no doubt exist, but in no spot upon the globe that I am yet acquainted with, do so many of them occur together, or do they form so decided a feature in the natural history of the country.

The Mammalia of Celebes are very few in number, consisting of fourteen terrestrial species and seven bats. Of the former no less than eleven are peculiar, including two which there is reason to believe may have been recently carried into other islands by man. Three species which have a tolerably wide range in the Archipelago, are: (1) The curious Lemur, Tarsius spectrum, which is found in all the islands as far westward as Malacca; (2) the common Malay Civet, Viverra tangalunga, which has a still wider range; and (3) a

Deer, which seems to be the same as the Rusa hippelaphus of Java, and was probably introduced by man at an early period.

The more characteristic species are as follow:

Cynopithecus nigrescens, a curious baboon-like monkey if not a true baboon, which abounds all over Celebes, and is found nowhere else but in the one small island of Batchian, into which it has probably been introduced accidentally. An allied species is found in the Philippines, but in no other island of the Archipelago is there anything resembling them. These creatures are about the size of a spaniel, of a jet-black colour, and have the projecting dog-like muzzle and overhanging brows of the baboons. They have large red callosities and a short fleshy tail, scarcely an inch long and hardly visible. They go in large bands, living chiefly in the trees, but often descending on the ground and robbing gardens and orchards.

Anoa depressicornis, the Sapi-utan, or wild cow of the Malays, is an animal which has been the cause of much controversy, as to whether it should be classed as ox, buffalo, or antelope. It is smaller than any other wild cattle, and in many respects seems to approach some of the ox-like antelopes of Africa. It is found only in the mountains, and is said never to inhabit places where there are deer. It is somewhat smaller than a small Highland cow, and has long straight horns, which are ringed at the base and slope backwards over the neck.

The wild pig seems to be of a species peculiar to the

island; but a much more curious animal of this family is the Babirusa or Pig-deer; so named by the Malays from its long and slender legs, and curved tusks resembling horns. This extraordinary creature resembles a pig in general appearance, but it does not dig with its snout, as it feeds on fallen fruits. The tusks of the lower jaw are very long and sharp, but the upper ones instead of growing downwards in the usual way are completely reversed, growing upwards out of bony sockets through the skin on each side of the snout, curving backwards to near the eyes, and in old animals often reaching eight or ten inches in length. It is difficult to understand what can be the use of these extraordinary horn-like teeth. Some of the old writers supposed that they served as hooks, by which the creature could rest its head on a branch. But the way in which they usually diverge just over and in front of the eye has suggested the more probable idea, that they serve to guard these organs from thorns and spines, while hunting for fallen fruits among the tangled thickets of rattans and other spiny plants. Even this, however, is not satisfactory, for the female, who must seek her food in the same way, does not possess them. I should be inclined to believe rather, that these tusks were once useful, and were then worn down as fast as they grew; but that changed conditions of life have rendered them unnecessary, and they now develop into a monstrous form, just as the incisors of the Beaver or Rabbit will go on growing, if the opposite teeth do not wear them away. In old animals they reach an enormous size, and are generally broken off as if by fighting.

Here again we have a resemblance to the Wart-hogs of Africa, whose upper canines grow outwards and curve up so as to form a transition from the usual mode of growth to that of the Babirusa. In other respects there seems no affinity between these animals, and the Babirusa stands completely isolated, having no resemblance to the pigs of any other part of the world. It is found all over Celebes and in the Sula islands, and also in Bourn, the only spot beyond the Celebes group to which it extends; and which island also shows some affinity to the Sula islands in its birds, indicating perhaps, a closer connection between them at some former period than now exists.

The other terrestrial mammals of Celebes are five species of squirrels, which are all distinct from those of Java and Borneo, and mark the furthest eastward range of the genus in the tropics; and two of Eastern opossums (Cuscus), which are different from those of the Moluccas, and mark the furthest westward extension of this genus and of the Marsupial order. Thus we see that the Mammalia of Celebes are no less individual and remarkable than the birds, since three of the largest and most interesting species have no near allies in surrounding countries, but seem vaguely to indicate a relation to the African continent.

Many groups of insects appear to be especially subject to

local influences, their forms and colours changing with each change of conditions, or even with a change of locality where the conditions seem almost identical. We should therefore anticipate that the individuality manifested in the higher animals would be still more prominent in these creatures with less stable organisms. On the other hand, however, we have to consider that the dispersion and migration of insects is much more easily effected than that of mammals or even of birds. They are much more likely to be carried away by violent winds; their eggs may be carried on leaves either by storms of wind or by floating trees, and their larvae and pupae, often buried in trunks of trees or enclosed in waterproof cocoons, may be floated for days or weeks uninjured over the ocean. These facilities of distribution tend to assimilate the productions of adjacent lands in two ways: first, by direct mutual interchange of species; and secondly, by repeated immigrations of fresh individuals of a species common to other islands, which by intercrossing, tend to obliterate the changes of form and colour, which differences of conditions might otherwise produce. Bearing these facts in mind, we shall find that the individuality of the insects of Celebes is even greater than we have any reason to expect.

For the purpose of insuring accuracy in comparisons with other islands, I shall confine myself to those groups which are best known, or which I have myself carefully studied. Beginning with the Papilionidae or Swallow-tailed butterflies, Celebes possesses 24 species, of which the large number of 18 are not found in any other island. If we compare this with Borneo, which out of 29 species has only two not found elsewhere, the difference is as striking as anything can be. In the family of the Pieridae, or white butterflies, the difference is not quite so great, owing perhaps to the more wandering habits of the group; but it is still very remarkable. Out of 30 species inhabiting Celebes, 19 are peculiar, while Java (from which more species are known than from Sumatra or Borneo), out of 37 species, has only 13 peculiar. The Danaidae are large, but weak-flying butterflies, which frequent forests and gardens, and are plainly but often very richly coloured. Of these my own collection contains 16 species from Celebes and 15 from Borneo; but whereas no less than 14 are confined to the former island, only two are peculiar to the latter. The Nymphalidae are a very extensive group, of generally strong-winged and very bright-coloured butterflies, very abundant in the tropics, and represented in our own country by our Fritillaries, our Vanessas, and our Purple-emperor. Some months ago I drew up a list of the Eastern species of this group, including all the new ones discovered by myself, and arrived at the following comparative results:—

Species of Nymphalidae.	Species peculiar to each island.	Percentage of peculiar

Volume I

			Species.
Java.....	70......	23..........	33
Borneo....	52......	15..........	29
Celebes ...	48......	35..........	73

The Coleoptera are so extensive that few of the groups have yet been carefully worked out. I will therefore refer to one only, which I have myself recently studied—the Cetoniadae or Rose-chafers—a group of beetles which, owing to their extreme beauty, have been much sought after. From Java 37 species of these insects are known, and from Celebes only 30; yet only 13, or 35 percent, are peculiar to the former island, and 19, or 63 percent, to the latter.

The result of these comparisons is, that although Celebes is a single, large island with only a few smaller ones closely grouped around it, we must really consider it as forming one of the great divisions of the Archipelago, equal in rank and importance to the whole of the Moluccan or Philippine groups, to the Papuan islands, or to the Indo-Malay islands (Java, Sumatra, Borneo, and the Malay peninsula). Taking those families of insects and birds which are best known, the following table shows the comparison of Celebes with the other groups of islands:—

	PAPILIONIDAE AND PERIDAE	HAWKS, PARROTS, AND PIGEONS.
	Percent of peculiar Species.	
Indo-Malay region....	56	54
Philippine group	66	73
Celebes.........	69	60
Moluccan group	52	62
Timor group.......	42	47
Papuan group	64	74

These large and well-known families well represent the general character of the zoology of Celebes; and they show that this island is really one of the most isolated portions of the Archipelago, although situated in its very centre.

But the insects of Celebes present us with other phenomena more curious and more difficult to explain than their striking individuality. The butterflies of that island are in many cases characterised by a peculiarity of outline, which distinguishes them at a glance from those of any other part of the world. It is most strongly manifested in the Papilios and the Pieridae, and consists in the forewings being either strongly curved or abruptly bent near the base, or in the extremity being elongated and often somewhat hooked. Out of the 14 species of Papilio in Celebes, 13 exhibit this peculiarity in a greater or less degree, when compared with the most nearly allied species of the surrounding islands. Ten species of Pieridae have the same character, and in four or five of the Nymphalidae it is also very distinctly marked. In almost every case, the species found in Celebes are much larger than those of the islands westward, and at least equal to those of the Moluccas, or even larger. The difference of form is, however, the most remarkable feature, as it is altogether a new thing for a whole set of species in one country to differ in exactly the same way from the corresponding sets in all the surrounding countries; and it is so well marked, that without looking at the details of colouring, most Celebes Papilios and many Pieridae, can be at once distinguished from those of other islands by their form alone.

The outside figure of each pair here given, shows the exact size and form of the fore-wing in a butterfly of Celebes, while the inner one represents the most closely allied species from one of the adjacent islands. Figure 1 shows the strongly curved margin of the Celebes species, Papilio gigon, compared with the much straighter margin of Papilio demolion from Singapore and Java. Figure 2 shows the abrupt bend over the base of the wing in Papilio miletus of Celebes, compared with the slight curvature in the common Papilio sarpedon, which has almost exactly the same form from India to New Guinea and Australia. Figure 3 shows the elongated wing of Tachyris zarinda, a native of

Celebes, compared with the much shorter wing of Tachyris nero, a very closely allied species found in all the western islands. The difference of form is in each case sufficiently obvious, but when the insects themselves are compared, it is much more striking than in these partial outlines.

From the analogy of birds, we should suppose that the pointed wing gave increased rapidity of flight, since it is a character of terns, swallows, falcons, and of the swift-flying pigeons. A short and rounded wing, on the other hand, always accompanies a more feeble or more laborious flight, and one much less under command. We might suppose, therefore, that the butterflies which possess this peculiar form were better able to escape pursuit. But there seems no unusual abundance of insectivorous birds to render this necessary; and as we cannot believe that such a curious peculiarity is without meaning, it seems probable that it is the result of a former condition of things, when the island possessed a much richer fauna, the relics of which we see in the isolated birds and Mammalia now inhabiting it; and when the abundance of insectivorous creatures rendered some unusual means of escape a necessity for the large-winged and showy butterflies. It is some confirmation of this view, that neither the very small nor the very obscurely coloured groups of butterflies have elongated wings, nor is any modification perceptible in those strong-winged groups which already possess great strength and rapidity of flight.

These were already sufficiently protected from their enemies, and did not require increased power of escaping from them. It is not at all clear what effect the peculiar curvature of the wings has in modifying flight.

Another curious feature in the zoology of Celebes is also worthy of attention. I allude to the absence of several groups which are found on both sides of it, in the Indo-Malay islands as well as in the Moluccas; and which thus seem to be unable, from some unknown cause, to obtain a footing in the intervening island. In Birds we have the two families of Podargidae and Laniadae, which range over the whole Archipelago and into Australia, and which yet have no representative in Celebes. The genera Ceyx among Kingfishers, Criniger among Thrushes, Rhipidura among Flycatchers, Calornis among Starlings, and Erythrura among Finches, are all found in the Moluccas as well as in Borneo and Java—but not a single species belonging to any one of them is found in Celebes. Among insects, the large genus of Rose-chafers, Lomaptera, is found in every country and island between India and New Guinea, except Celebes. This unexpected absence of many groups, from one limited district in the very centre of their area of distribution, is a phenomenon not altogether unique, but, I believe, nowhere so well marked as in this case; and it certainly adds considerably to the strange character of this remarkable island.

The anomalies and eccentricities in the natural history of Celebes which I have endeavoured to sketch in this CHAPTER, all point to an origin in a remote antiquity. The history of extinct animals teaches us that their distribution in time and in space are strikingly similar. The rule is, that just as the productions of adjacent areas usually resemble each other closely, so do the productions of successive periods in the same area; and as the productions of remote areas generally differ widely, so do the productions of the same area at remote epochs. We are therefore led irresistibly to the conclusion, that change of species, still more of generic and of family form, is a matter of time. But time may have led to a change of species in one country, while in another the forms have been more permanent, or the change may have gone on at an equal rate but in a different manner in both. In either case, the amount of individuality in the productions of a district will be to some extent a measure of the time that a district has been isolated from those that surround it. Judged by this standard, Celebes must be one of the oldest parts of the Archipelago. It probably dates from a period not only anterior to that when Borneo, Java, and Sumatra were separated from the continent, but from that still more remote epoch when the land that now constitutes these islands had not risen above the ocean.

Such an antiquity is necessary, to account for the

number of animal forms it possesses, which show no relation to those of India or Australia, but rather with those of Africa; and we are led to speculate on the possibility of there having once existed a continent in the Indian Ocean which might serve as a bridge to connect these distant countries. Now it is a curious fact, that the existence of such a land has been already thought necessary, to account for the distribution of the curious Quadrumana forming the family of the Lemurs. These have their metropolis in Madagascar, but are found also in Africa, in Ceylon, in the peninsula of India, and in the Malay Archipelago as far as Celebes, which is its furthest eastern limit. Dr. Sclater has proposed for the hypothetical continent connecting these distant points, and whose former existence is indicated by the Mascarene islands and the Maldive coral group, the name of Lemuria. Whether or not we believe in its existence in the exact form here indicated, the student of geographical distribution must see in the extraordinary and isolated productions of Celebes, proof of the former existence of some continent from whence the ancestors of these creatures, and of many other intermediate forms, could have been derived.

In this short sketch of the most striking peculiarities of the Natural History of Celebes, I have been obliged to enter much into details that I fear will have been uninteresting to the general reader, but unless I had done so, my exposition would have lost much of its force and value. It is by these details alone that I have been able to prove the unusual features that Celebes presents to us. Situated in the very midst of an Archipelago, and closely hemmed in on every side by islands teeming with varied forms of life, its productions have yet a surprising amount of individuality. While it is poor in the actual number of its species, it is yet wonderfully rich in peculiar forms, many of which are singular or beautiful, and are in some cases absolutely unique upon the globe. We behold here the curious phenomenon of groups of insects changing their outline in a similar manner when compared with those of surrounding islands, suggesting some common cause which never seems to have acted elsewhere in exactly the same way. Celebes, therefore, presents us with a most striking example of the interest that attaches to the study of the geographical distribution of animals. We can see that their present distribution upon the globe is the result of all the more recent changes the earth's surface has undergone; and, by a careful study of the phenomena, we are sometimes able to deduce approximately what those past changes must have been in order to produce the distribution we find to exist. In the comparatively simple case of the Timor group, we were able to deduce these changes with some approach to certainty. In the much more complicated case of Celebes, we can only indicate their general nature, since we now see the result, not of any single or recent change only, but of a whole series of the later revolutions which have resulted in the present distribution of land in the Eastern Hemisphere.

CHAPTER XIX. BANDA.

(DECEMBER 1857, MAY 1859, APRIL 1861.)

THE Dutch mail steamer in which I travelled from Macassar to Banda and Amboyna was a roomy and comfortable vessel, although it would only go six miles an hour in the finest weather. As there were but three passengers besides myself, we had abundance of room, and I was able to enjoy a voyage more than I had ever done before. The arrangements are somewhat different from those on board English or Indian steamers. There are no cabin servants, as every cabin passenger invariably brings his own, and the ship's stewards attend only to the saloon and the eating department. At six A.M. a cup of tea or coffee is provided for those who like it. At seven to eight there is a light breakfast of tea, eggs, sardines, etc. At ten, Madeira, Gin and bitters are brought on deck as a whet for the substantial eleven o'clock breakfast, which differs from a dinner only in the absence of soup. Cups of tea and coffee are brought around at three P.M.; bitters, etc. again at five, a good dinner with beer and claret at half-past six, concluded by tea and coffee at eight. Between whiles, beer and sodawater are supplied when called for, so there is no lack of little gastronomical excitements to while away the tedium of a sea voyage.

Our first stopping place was Coupang, at the west end of the large island of Timor. We then coasted along that island for several hundred miles, having always a view of hilly ranges covered with scanty vegetation, rising ridge behind ridge to the height of six or seven thousand feet. Turning off towards Banda we passed Pulo-Cambing, Wetter, and Roma, all of which are desolate and barren volcanic islands, almost as uninviting as Aden, and offering a strange contrast to the usual verdure and luxuriance of the Archipelago. In two days more we reached the volcanic group of Banda, covered with an unusually dense and brilliant green vegetation, indicating that we had passed beyond the range of the hot dry winds from the plains of Central Australia. Banda is a lovely little spot, its three islands enclosing a secure harbour from whence no outlet is visible, and with water so transparent, that living corals and even the minutest objects are plainly seen on the volcanic sand at a depth of seven or eight fathoms. The ever smoking volcano rears its bare cone on one side, while the two larger islands are clothed with vegetation to the summit of the hills.

Going on shore, I walked up a pretty path which leads to the highest point of the island on which the town is situated, where there is a telegraph station and a magnificent view. Below lies the little town, with its neat red-tiled white houses and the thatched cottages of the natives, bounded on one side by the old Portuguese fort. Beyond, about half a mile distant, lies the larger island in the shape of a horseshoe, formed of a range of abrupt hills covered with fine forest and nutmeg gardens; while close opposite the town is the volcano, forming a nearly perfect cone, the lower part only covered with a light green bushy vegetation. On its north side the outline is more uneven, and there is a slight hollow or chasm about one-fifth of the way down, from which constantly issue two columns of smoke, as well as a good deal from the rugged surface around and from some spots nearer the summit. A white efflorescence, probably sulphur, is thickly spread over the upper part of the mountain, marked by the narrow black vertical lines of water gullies. The smoke unites as it rises, and forms a dense cloud, which in calm, damp weather spreads out into a wide canopy hiding the top of the mountain. At night and early morning, it often rises up straight and leaves the whole outline clear.

It is only when actually gazing on an active volcano that one can fully realize its awfulness and grandeur. Whence comes that inexhaustible fire whose dense and sulphurous smoke forever issues from this bare and desolate peak? Whence the mighty forces that produced that peak, and still from time to time exhibit themselves in the earthquakes that always occur in the vicinity of volcanic vents? The knowledge from childhood of the fact that volcanoes and earthquakes exist, has taken away somewhat of the strange and exceptional character that really belongs to them. The inhabitant of most parts of northern Europe sees in the earth the emblem of stability and repose. His whole life-experience, and that of all his age and generation, teaches him that the earth is solid and firm, that its massive rocks may contain water in abundance, but never fire; and these essential characteristics of the earth are manifest in every mountain his country contains. A volcano is a fact opposed to all this mass of experience, a fact of so awful a character that, if it were the rule instead of the exception, it would make the earth uninhabitable a fact so strange and unaccountable that we may be sure it would not be believed on any human testimony, if presented to us now for the first time, as a natural phenomenon happening in a distant country.

The summit of the small island is composed of a highly crystalline basalt; lower down I found a hard, stratified slatey sandstone, while on the beach are huge blocks of lava, and scattered masses of white coralline limestone. The larger island has coral rock to a height of three or four hundred feet, while above is lava and basalt. It seems probable, therefore, that this little group of four islands is the fragment of a larger district which was perhaps once connected with Ceram, but which was separated and broken up by the same forces which formed the volcanic cone. When I visited the larger island on another occasion, I saw a considerable tract covered with large forest trees—dead, but still standing. This was a record of the last great earthquake only two years ago, when the sea broke in over this part of the island and so flooded it as to destroy the vegetation on all the lowlands. Almost every year there is an earthquake here, and at intervals of a few years, very severe ones which throw down houses and carry ships out of the harbour bodily into the streets.

Notwithstanding the losses incurred by these terrific visitations, and the small size and isolated position of these little islands, they have been and still are of considerable value to the Dutch Government, as the chief nutmeg-garden in the world. Almost the whole surface is planted with nutmegs, grown under the shade of lofty Kanary trees (Kanarium commune). The light volcanic soil, the shade, and the excessive moisture of these islands, where it rains more or less every month in the year, seem exactly to suit the nutmeg-tree, which requires no manure and scarcely any attention. All the year round flowers and ripe fruit are to be found, and none of those diseases occur which under a forced and unnatural system of cultivation have ruined the nutmeg planters of Singapore and Penang.

Few cultivated plants are more beautiful than nutmeg-trees. They are handsomely shaped and glossy-leaved, growing to the height of twenty or thirty feet, and bearing small yellowish flowers. The fruit is the size and colour of a peach, but rather oval. It is of a tough fleshy consistence, but when ripe splits open, and shows the dark-brown nut within, covered with the crimson mace, and is then a most beautiful object. Within the thin, hard shell of the nut is the seed, which is the nutmeg of commerce. The nuts are eaten by the large pigeons of Banda, which digest the mace, but cast up the nut with its seed uninjured.

The nutmeg trade has hitherto been a strict monopoly of the Dutch Government; but since leaving the country I believe that this monopoly has been partially or wholly discontinued, a proceeding which appears exceedingly injudicious and quite unnecessary. There are cases in which monopolies are perfectly justifiable, and I believe this to be one of them. A small country like Holland cannot afford to keep distant and expensive colonies at a loss; and having possession of a very small island where a valuable product, not a necessity of life, can be obtained at little cost, it is almost the duty of the state to monopolise it. No injury is done thereby to anyone, but a great benefit is conferred upon the whole population of Holland and its dependencies, since the produce of the state monopolies saves them from the weight of a heavy taxation. Had the Government not kept the nutmeg trade of Banda in its own hands, it is probable that the whole of the islands would long ago have become the property of one or more large capitalists. The monopoly would have been almost the same, since no known spot on the globe can produce nutmegs so cheaply as Banda, but the profits of the monopoly would have gone to a few individuals instead of to the nation.

As an illustration of how a state monopoly may become a state duty, let us suppose that no gold existed in Australia, but that it had been found in immense quantities by one of our ships in some small and barren island. In this case it would plainly become the duty of the state to keep and work the mines for the public benefit, since by doing so, the gain would be fairly divided among the whole population by decrease of taxation; whereas by leaving it open to free trade while merely keeping the government of the island; we should certainly produce enormous evils during the first struggle for the precious metal, and should ultimately subside into the monopoly of some wealthy individual or great company, whose enormous revenue would not equally benefit the community. The nutmegs of Banda and the tin of Banca are to some extent parallel cases to this supposititious one, and I believe the Dutch Government will act most unwisely if they give up their monopoly.

Even the destruction of the nutmeg and clove trees in many islands, in order to restrict their cultivation to one or two where the monopoly could be easily guarded, usually made the theme of so much virtuous indignation against the Dutch, may be defended on similar principles, and is certainly not nearly so bad as many monopolies we ourselves have until very recently maintained. Nutmegs and cloves are not necessaries of life; they are not even used as spices by the natives of the Moluccas, and no one was materially or permanently injured by the destruction of the trees, since there are a hundred other products that can be grown in the same islands, equally valuable and far more beneficial in a social point of view. It is a case exactly parallel to our prohibition of the growth of tobacco in England, for fiscal purposes, and is, morally and economically, neither better nor worse. The salt monopoly which we so long maintained in India was in much worse. As long as we keep up a system of excise and customs on articles of daily use, which requires an elaborate array of officers and coastguards to carry into effect, and which creates a number of purely legal crimes, it is the height of absurdity for us to affect indignation at the conduct of the Dutch, who carried out a much more justifiable, less hurtful, and more profitable system in their Eastern possessions.

I challenge objectors to point out any physical or moral evils that have actually resulted from the action of the Dutch Government in this matter; whereas such evils are the admitted results of every one of our monopolies and restrictions. The conditions of the two experiments are totally different. The true "political economy" of a higher race, when governing a lower race, has never yet been worked out. The application of our "political economy" to such cases invariably results in the extinction or degradation of the lower race; whence, we may consider it probable that one of the necessary conditions of its truth is the approximate mental and social unity of the society in which it is applied. I shall again refer to this subject in my CHAPTER on Ternate, one of the most celebrated of the old spice-islands.

The natives of Banda are very much mixed, and it is probable that at least three-fourths of the population are mongrels, in various degrees of Malay, Papuan, Arab, Portuguese, and Dutch. The first two form the bases of the larger portion, and the dark skins, pronounced features, and more or less frizzly hair of the Papuans preponderates. There seems little doubt that the aborigines of Banda were Papuans, and a portion of them still exists in the Ke islands, where they emigrated when the Portuguese first took possession of their native island. It is such people as these that are often looked upon as transitional forms between two very distinct races, like the Malays and Papuans, whereas they are only examples of intermixture.

The animal productions of Banda, though very few, are interesting. The islands have perhaps no truly indigenous Mammalia but bats. The deer of the Moluccas and the pig have probably been introduced. A species of Cuscus or Eastern opossum is also found at Banda, and this may be truly indigenous in the sense of not having been introduced by man. Of birds, during my three visits of one or two days each, I collected eight kinds, and the Dutch collectors have added a few others. The most remarkable is a fine and very handsome fruit-pigeon, *Carpophaga concinna*, which feeds upon the nutmegs, or rather on the mace, and whose loud booming note is to be continually heard. This bird is found in the Ke and Matabello islands as well as Banda, but not in Ceram or any of the larger islands, which are inhabited by allied but very distinct species. A beautiful small fruit-dove, Ptilonopus diadematus, is also peculiar to Banda.

CHAPTER XX. AMBOYNA.

(DECEMBER 1857, OCTOBER 1859, FEBRUARY 1860.)

TWENTY hours from Banda brought us to Amboyna, the capital of the Moluccas, and one of the oldest European settlements in the East. The island consists of two peninsulas, so nearly divided by inlets of the sea, as to leave only a sandy isthmus about a mile wide near their eastern extremity. The western inlet is several miles long and forms a fine harbour on the southern side of which is situated the town of Amboyna. I had a letter of introduction to Dr. Mohnike, the chief medical officer of the Moluccas, a German and a naturalist. I found that he could write and read English, but could not speak it, being like myself a bad linguist; so we had to use French as a medium of communication. He kindly offered me a room during my stay in Amboyna, and introduced me to his junior, Dr. Doleschall, a Hungarian and also an entomologist. He was an intelligent and most amiable young man but I was shocked to find that he was dying of consumption, though still able to perform the duties of his office. In the evening my host took me to the residence of the Governor, Mr. Goldmann, who received me in a most kind and cordial manner, and offered me every assistance. The town of Amboyna consists of a few business streets, and a number of roads set out at right angles to each other, bordered by hedges of flowering shrubs, and enclosing country houses and huts embossed in palms and fruit trees. Hills and mountains form the background in almost every direction, and there are few places more enjoyable for a morning or evening stroll than these sandy roads and shady lanes in the suburbs of the ancient city of Amboyna.

There are no active volcanoes in the island, nor is it now subject to frequent earthquakes, although very severe ones have occurred and may be expected again. Mr. William Funnell, in his voyage with Dampier to the South Seas in 1705, says: "Whilst we were here, (at Amboyna) we had a great earthquake, which continued two days, in which time it did a great deal of mischief, for the ground burst open in many places, and swallowed up several houses and whole families. Several of the people were dug out again, but most of them dead, and many had their legs or arms broken by the fall of the houses. The castle walls were rent asunder in several places, and we thought that it and all the houses would have fallen down. The ground where we were swelled like a wave in the sea, but near us we had no hurt done." There are also numerous records of eruptions of a volcano on the west side of the island. In 1674 an eruption destroyed a village. In 1694 there was another eruption. In 1797 much vapour and heat was emitted. Other eruptions occurred in

1816 and 1820, and in 1824 a new crater is said to have been formed. Yet so capricious is the action of these subterranean fires, that since the last-named epoch all eruptive symptoms have so completely ceased, that I was assured by many of the most intelligent European inhabitants of Amboyna, that they had never heard of any such thing as a volcano on the island.

During the few days that elapsed before I could make arrangements to visit the interior, I enjoyed myself much in the society of the two doctors, both amiable and well-educated men, and both enthusiastic entomologists, though obliged to increase their collections almost entirely by means of native collectors. Dr. Doleschall studied chiefly the flies and spiders, but also collected butterflies and moths, and in his boxes I saw grand specimens of the emerald Ornithoptera priamus and the azure Papilio ulysses, with many more of the superb butterflies of this rich island. Dr. Mohnike confined himself chiefly to the beetles, and had formed a magnificent collection during many years residence in Java, Sumatra, Borneo, Japan, and Amboyna. The Japanese collection was especially interesting, containing both the fine Carabi of northern countries, and the gorgeous Buprestidae and Longicorns of the tropics. The doctor made the voyage to Jeddo by land from Nagasaki, and is well acquainted with the character, manners, and customs of the people of Japan, and with the geology, physical features, and natural history of the country. He showed me collections of cheap woodcuts printed in colours, which are sold at less than a farthing each, and comprise an endless variety of sketches of Japanese scenery and manners. Though rude, they are very characteristic, and often exhibit touches of great humour. He also possesses a large collection of coloured sketches of the plants of Japan, made by a Japanese lady, which are the most masterly things I have ever seen. Every stem, twig, and leaf is produced by single touches of the brush, the character and perspective of very complicated plants being admirably given, and the articulations of stem and leaves shown in a most scientific manner.

Having made arrangements to stay for three weeks at a small hut on a newly cleared plantation in the interior of the northern half of the island, I with some difficulty obtained a boat and men to take me across the water—for the Amboynese are dreadfully lazy. Passing up the harbour, in appearance like a fine river, the clearness of the water afforded me one of the most astonishing and beautiful sights I have ever beheld. The bottom was absolutely hidden by a continuous series of corals, sponges, actiniae, and other marine productions of magnificent dimensions, varied forms, and brilliant colours. The depth varied from about twenty to fifty feet, and the bottom was very uneven, rocks and chasms and little hills and valleys, offering a variety of stations for the growth of these animal forests. In and out among them, moved numbers of blue and red and yellow fishes, spotted and banded and striped in the most striking manner, while great orange or rosy transparent medusae floated along near the surface. It was a sight to gaze at for hours, and no description can do justice to its surpassing beauty and interest. For once, the reality exceeded the most glowing accounts I had ever read of the wonders of a coral sea. There is perhaps no spot in the world richer in marine productions, corals, shells and fishes, than the harbour of Amboyna.

From the north side of the harbour, a good broad path passes through swamp, clearing and forest, over hill and valley, to the farther side of the island; the coralline rock constantly protruding through the deep red earth which fills all the hollows, and is more or less spread over the plains and hill-sides. The forest vegetation is here of the most luxuriant character; ferns and palms abound, and the climbing rattans were more abundant than I had ever seen them, forming tangled festoons over almost every large forest tree. The cottage I was to occupy was situated in a large clearing of about a hundred acres, part of which was already planted with young cacao-trees and plantains to shade them, while the rest was covered with dead and half-burned forest trees; and on one side there was a tract where the trees had been recently felled and were not yet burned. The path by which I had arrived continued along one side of this clearing, and then again entering the virgin forest passed over hill and dale to the northern aide of the island.

My abode was merely a little thatched hut, consisting of an open verandah in front and a small dark sleeping room behind. It was raised about five feet from the ground, and was reached by rude steps to the centre of the verandah. The walls and floor were of bamboo, and it contained a table, two bamboo chairs, and a couch. Here I soon made myself comfortable, and set to work hunting for insects among the more recently felled timber, which swarmed with fine Curculionidae, Longicorns, and Buprestidae, most of them remarkable for their elegant forms or brilliant colours, and almost all entirely new to me. Only the entomologist can appreciate the delight with which I hunted about for hours in the hot sunshine, among the branches and twigs and bark of the fallen trees, every few minutes securing insects which were at that time almost all rare or new to European collections.

In the shady forest paths were many fine butterflies, most conspicuous among which was the shining blue Papilio ulysses, one of the princes of the tribe, though at that time so rare in Europe, I found it absolutely common in Amboyna, though not easy to obtain in fine condition, a large number of the specimens being found when captured to have the wings torn or broken. It flies with a rather weak undulating motion, and from its large size, its tailed wings and brilliant colour, is one of the most tropical-looking insects the naturalist can gaze upon.

There is a remarkable contrast between the beetles of Amboyna and those of Macassar, the latter generally small and obscure, the former large and brilliant. On the whole, the insects here most resemble those of the Aru islands, but they are almost always of distinct species, and when they are most nearly allied to each other, the species of Amboyna are of larger size and more brilliant colours, so that one might be led to conclude that in passing east and west into a less favourable soil and climate, they had degenerated into less striking forms.

Of an evening I generally sat reading in the verandah, ready to capture any insects that were attracted to the light. One night about nine o'clock, I heard a curious noise and rustling overhead, as if some heavy animal were crawling slowly over the thatch. The noise soon ceased, and I thought no more about it and went to bed soon afterwards. The next afternoon just before dinner, being rather tired with my day's work, I was lying on the couch with a book in my hand, when gazing upwards I saw a large mass of something overhead which I had not noticed before. Looking more carefully I could see yellow and black marks, and thought it must be a tortoise-shell put up there out of the way between the ridge-pole and the roof. Continuing to gaze, it suddenly resolved itself into a large snake, compactly coiled up in a kind of knot; and I could detect his head and his bright eyes in the very centre of the folds. The noise of the evening before was now explained. A python had climbed up one of the posts of the house, and had made his way under the thatch within a yard of my head, and taken up a comfortable position in the roof—and I had slept soundly all night directly under him.

I called to my two boys who were skinning birds below and said, "Here's a big snake in the roof;" but as soon as I had shown it to them they rushed out of the house and begged me to come out directly. Finding they were too much afraid to do anything, we called some of the labourers in the plantation, and soon had half a dozen men in consultation outside. One of these, a native of Bouru, where there are a great many snakes, said he would get him out, and proceeded to work in a businesslike manner. He made a strong noose of rattan, and with a long pole in the other hand poked at the snake, who then began slowly to uncoil itself. He then managed to slip the noose over its head, and getting it well on to the body, dragged the animal down. There was a great scuffle as the snake coiled round the chairs and posts to resist his enemy, but at length the man caught hold of its tail, rushed out of the house (running so quick that the creature seemed quite confounded), and tried to strike its head against a tree. He missed however, and let go, and the snake got under a dead trunk close by. It was again poked out, and again the Bouru man caught hold of its tail, and running away quickly dashed its head with a swing against a tree, and it was then easily killed with a hatchet. It was about twelve feet long and very thick, capable of doing much mischief and of swallowing a dog or a child.

I did not get a great many birds here. The most remarkable were the fine crimson lory, Eos rubra—a brush-tongued parroquet of a vivid crimson colour, which was very abundant. Large flocks of them came about the plantation, and formed a magnificent object when they settled down upon some flowering tree, on the nectar of which lories feed. I also obtained one or two specimens of the fine racquet-tailed kingfisher of Amboyna, Tanysiptera nais, one of the most singular and beautiful of that beautiful family. These birds differ from all other kingfishers (which have usually short tails) by having the two middle tail-feathers immensely lengthened and very narrowly webbed, but terminated by a spoon-shaped enlargement, as in the motmots and some of the humming-birds. They belong to that division of the family termed king-hunters, living chiefly on insects and small land-molluscs, which they dart down upon and pick up from the ground, just as a kingfisher picks a fish out of the water. They are confined to a very limited area, comprising the Moluccas, New Guinea and Northern Australia. About ten species of these birds are now known, all much resembling each other, but yet sufficiently distinguishable in every locality. The Amboynese species, of which a very accurate representation is here given, is one of the largest and handsomest. It is full seventeen inches long to the tips of the tail-feathers; the bill is coral red, the under-surface pure white, the back and wings deep purple, while the shoulders, head and nape, and some spots on the upper part of the back and wings, are pure azure blue; the tail is white, with the feathers narrowly blue-edged, but the narrow part of the long feathers is rich blue. This was an entirely new species, and has been well named after an ocean goddess, by Mr. R. G. Gray.

On Christmas eve I returned to Amboyna, where I stayed about ten days with my kind friend Dr. Mohnike. Considering that I had been away only twenty days, and that on five or six of those I was prevented doing anything by wet weather and slight attacks of fever, I had made a very nice collection of insects, comprising a much larger proportion of large and brilliant species than I had ever before obtained in so short a time. Of the beautiful metallic Buprestidae I had about a dozen handsome species, yet in the doctor's collection I observed four or five more very fine ones, so that Amboyna is unusually rich in this elegant group.

During my stay here I had a good opportunity of seeing how Europeans live in the Dutch colonies, and where they have adopted customs far more in accordance with the climate than we have done in our tropical possessions. Almost all business is transacted in the morning between the hours of seven and twelve, the afternoon being given up to repose, and the evening to visiting. When in the house during the heat of the day, and even at dinner, they use a loose cotton dress, only putting on a suit of thin European-made clothes for out of doors and evening wear. They often walk about after sunset bareheaded, reserving the black hat for visits of ceremony. Life is thus made far more agreeable, and the fatigue and discomfort incident to the climate greatly diminished. Christmas day is not made much of, but on New Year's day official and complimentary visits are paid, and about sunset we went to the Governor's, where a large party of ladies and gentlemen were assembled. Tea and coffee were handed around, as is almost universal during a visit, as well as cigars, for on no occasion is smoking prohibited in Dutch colonies, cigars being generally lighted before the cloth is withdrawn at dinner, even though half the company are ladies. I here saw for the first time the rare black lory from New Guinea, Chalcopsitta atra. The plumage is rather glossy, and slightly tinged with yellowish and purple, the bill and feet being entirely black.

The native Amboynese who reside in the city are a strange half-civilized, half-savage lazy people, who seem to be a mixture of at least three races—Portuguese, Malay, and Papuan or Ceramese, with an occasional cross of Chinese or Dutch. The Portuguese element decidedly predominates in the old Christian population, as indicated by features, habits, and the retention of many Portuguese words in the Malay, which is now their language. They have a peculiar style of dress which they wear among themselves, a close-fitting white shirt with black trousers, and a black frock or upper shirt. The women seem to prefer a dress entirely black. On festivals and state occasions they adopt the swallow-tail coat, chimneypot hat, and their accompaniments, displaying all the absurdity of our European fashionable dress. Though now Protestants, they preserve at feasts and weddings the processions and music of the Catholic Church, curiously mixed up with the gongs and dances of the aborigines of the country. Their language has still much more Portuguese than Dutch in it, although they have been in close communication with the latter nation for more than two hundred and fifty years; even many names of birds, trees and other natural objects, as well as many domestic terms, being plainly Portuguese. [The following are a few of the Portuguese words in common use by the Malay-speaking natives of Amboyna and the other Molucca islands: Pombo (pigeon); milo (maize); testa (forehead); horas (hours); alfinete (pin); cadeira (chair); lenco (handkerchief); fresco (cool); trigo (flour); sono (sloop); familia (family); histori (talk); vosse (you); mesmo (even); cunhado (brother-in-law); senhor (sir); nyora for signora (madam). None of them, however, have the least notion that these words belong to a European language.] This people seems to have had a marvellous power of colonization, and a capacity for impressing their national characteristics on every country they conquered, or in which they effected a merely temporary settlement. In a suburb of Amboyna there is a village of aboriginal Malays who are Mahometans, and who speak a peculiar language allied to those of Ceram, as well as Malay. They are chiefly fishermen, and are said to be both more industrious and more honest than the native Christians.

I went on Sunday, by invitation, to see a collection of shells and fish made by a gentleman of Amboyna. The fishes are perhaps unrivalled for variety and beauty by those of any one spot on the earth. The celebrated Dutch ichthyologist, Dr. Blecker, has given a catalogue of seven hundred and eighty species found at Amboyna, a number almost equal to those of all the seas and rivers of Europe. A large proportion of them are of the most brilliant colours, being marked with bands and spots of the purest yellows, reds, and blues; while their forms present all that strange and endless variety so characteristic of the inhabitants of the ocean. The shells are also very numerous, and comprise a number of the finest species in the world. The Mactras and Ostreas in particular struck me by the variety and beauty of their colours. Shells have long been an object of traffic in Amboyna; many of the natives get their living by collecting and cleaning them, and almost every visitor takes away a small collection. The result is that many of the commoner-sorts have lost all value in the eyes of the amateur, numbers of the handsome but very common cones, cowries, and olives sold in the streets of London for a penny each, being natives of the distant isle of Amboyna, where they cannot be bought so cheaply. The fishes in the collection were all well preserved in clear spirit in hundreds of glass jars, and the shells were arranged in large shallow pith boxes lined with paper, every specimen being fastened down with thread. I roughly estimated that there were nearly a thousand different kinds of shells, and perhaps ten thousand specimens, while the collection of Amboyna fishes was nearly perfect.

On the 4th of January I left Amboyna for Ternate; but two years later, in October 1859, I again visited it after my residence in Menado, and stayed a month in the town in a small house which I hired for the sake of assorting and packing up a large and varied collection which I had brought with me from North Celebes, Ternate, and Gilolo. I was obliged to do this because the mail steamer would have come the following month by way of Amboyna to Ternate, and I should have been delayed two months before I could have reached the former place. I then paid my first visit to Ceram, and on returning to prepare for my second more complete exploration of that island, I stayed (much against my will) two months at Paso, on the isthmus which connects the two portions of the island of Amboyna. This village is situated on the eastern side of the isthmus, on sandy ground, with a very pleasant view over the sea to the island of Harúka. On the Amboyna side of the isthmus there is a small river which has been continued by a shallow canal to within thirty yards of high-water mark on the other side. Across this small space, which is sandy and but slightly elevated, all small boats and praus can be easily dragged, and all the smaller traffic from Ceram and the islands of Saparúa and Harúka, passes through Paso. The canal is not continued quite through, merely because every spring-tide would throw up just such a sand-bank as now exists.

I had been informed that the fine butterfly Ornithoptera priamus was plentiful here, as well as the racquet-tailed kingfisher and the ring-necked lory. I found, however, that I had missed the time for the former, and birds of all kinds were very scarce, although I obtained a few good ones, including one or two of the above-mentioned rarities. I was much pleased to get here the fine long-armed chafer, Euchirus longimanus. This extraordinary insect is rarely or never captured except when it comes to drink the sap of the sugar palms, where it is found by the natives when they go early in the morning to take away the bamboos which have been filled during the night. For some time one or two were brought me every day, generally alive. They are sluggish insects, and pull themselves lazily along by means of their immense forelegs. A figure of this and other Moluccan beetles is given in the 27th CHAPTER of this work.

Volume I

I was kept at Paso by an inflammatory eruption, brought on by the constant attacks of small acari-like harvest-bugs, for which the forests of Ceram are famous, and also by the want of nourishing food while in that island. At one time I was covered with severe boils. I had them on my eye, cheek, armpits, elbows, back, thighs, knees, and ankles, so that I was unable to sit or walk, and had great difficulty in finding a side to lie upon without pain. These continued for some weeks, fresh ones coming out as fast as others got well; but good living and sea baths ultimately cured them.

About the end of January Charles Allen, who had been my assistant in Malacca and Borneo, again joined me on agreement for three years; and as soon as I got tolerably well, we had plenty to do laying in stores and making arrangements for our ensuing campaign. Our greatest difficulty was in obtaining men, but at last we succeeded in getting two each. An Amboyna Christian named Theodorus Matakena, who had been some time with me and had learned to skin birds very well, agreed to go with Allen, as well as a very quiet and industrious lad named Cornelius, whom I had brought from Menado. I had two Amboynese, named Petrus Rehatta, and Mesach Matakena; the latter of whom had two brothers, named respectively Shadrach and Abednego, in accordance with the usual custom among these people of giving only Scripture names to their children.

During the time I resided in this place, I enjoyed a

luxury I have never met with either before or since—the true bread-fruit. A good deal of it has been planted about here and in the surrounding villages, and almost every day we had opportunities of purchasing some, as all the boats going to Amboyna were unloaded just opposite my door to be dragged across the isthmus. Though it grows in several other parts of the Archipelago, it is nowhere abundant, and the season for it only lasts a short time. It is baked entire in the hot embers, and the inside scooped out with a spoon. I compared it to Yorkshire pudding; Charles Allen said it was like mashed potatoes and milk. It is generally about the size of a melon, a little fibrous towards the centre, but everywhere else quite smooth and puddingy, something in consistence between yeast-dumplings and batter-pudding. We sometimes made curry or stew of it, or fried it in slices; but it is no way so good as simply baked. It may be eaten sweet or savory. With meat and gravy it is a vegetable superior to any I know, either in temperate or tropical countries. With sugar, milk, butter, or treacle, it is a delicious pudding, having a very slight and delicate but characteristic flavour, which, like that of good bread and potatoes, one never gets tired of. The reason why it is comparatively scarce is that it is a fruit of which the seeds are entirely aborted by cultivation, and the tree can therefore only be propagated by cuttings. The seed-bearing variety is common all over the tropics, and though the seeds are very good eating, resembling chestnuts,

Volume I

the fruit is quite worthless as a vegetable. Now that steam and Ward's cases render the transport of young plants so easy, it is much to be wished that the best varieties of this unequalled vegetable should be introduced into our West India islands, and largely propagated there. As the fruit will keep some time after being gathered, we might then be able to obtain this tropical luxury in Covent Garden Market.

Although the few months I at various times spent in Amboyna were not altogether very profitable to me in the way of collections, it will always remain as a bright spot in the review of my Eastern travels, since it was there that I first made the acquaintance of those glorious birds and insects which render the Moluccas classic ground in the eyes of the naturalist, and characterise its fauna as one of the most remarkable and beautiful upon the globe. On the 20th of February I finally quitted Amboyna for Ceram and Waigiou, leaving Charles Allen to go by a Government boat to Wahai on the north coast of Ceram, and thence to the unexplored island of Mysol.

www.ingramcontent.com/pod-product-compliance
Lightning Source LLC
Chambersburg PA
CBHW031323230426
43670CB00006B/222